MACMILLAN/McGRAW-HILL
Math

Daily Practice Workbook with Summer Skills Refresher

Grade 6

McGraw Hill

The McGraw·Hill Companies

**Macmillan
McGraw-Hill**

Published by Macmillan/McGraw-Hill, of McGraw-Hill Education, a division of The McGraw-Hill Companies, Inc., Two Penn Plaza, New York, New York 10121.

Copyright © by Macmillan/McGraw-Hill. All rights reserved. The contents, or parts thereof, may be reproduced in print form for non-profit educational use with Macmillan/McGraw-Hill Math, provided such reproductions bear copyright notice, but may not be reproduced in any form for any other purpose without the prior written consent of The McGraw-Hill Companies, Inc., including, but not limited to, network storage or transmission, or broadcast for distance learning.

Printed in the United States of America

3 4 5 6 7 8 9 024 08 07 06 05 04

Contents

Daily Practice

Lesson 1-1 Place Value 1
Lesson 1-2 Compare and Order 2
Lesson 1-3 Add and Subtract Whole Numbers and Decimals 3
Lesson 1-4 Estimate Sums and Differences 4
Lesson 1-5 Problem Solving: Skill Using the Four-Step Process 5

Lesson 2-1 Algebra: Explore Addition and Subtraction Expressions 6
Lesson 2-2 Problem Solving: Strategy Write an Equation 7
Lesson 2-3 Algebra: Properties of Addition 8
Lesson 2-4 Choose a Computation Method 9

Lesson 3-1 Multiplication Patterns 10
Lesson 3-2 Multiply Whole Numbers 11
Lesson 3-3 Exponents 12
Lesson 3-4 Estimate Products 13
Lesson 3-5 Problem Solving: Skill Estimate or Exact Answer 14
Lesson 3-6 Multiply Decimals 15
Lesson 3-7 Algebra: Properties and Mental Math 16

Lesson 4-1 Estimate Quotients 17
Lesson 4-2 Divide Whole Numbers 18
Lesson 4-3 Divide Decimals by Whole Numbers 19
Lesson 4-4 Algebra: Explore Multiplication and Division Expressions 20
Lesson 4-5 Problem Solving: Strategy Guess and Check 21
Lesson 4-6 Multiply and Divide by Powers of Ten 22
Lesson 4-7 Divide Decimals by Decimals 23

Lesson 5-1 Collect, Organize, and Display Data 24
Lesson 5-2 Bar Graphs 25
Lesson 5-3 Histograms 26
Lesson 5-4 Problem Solving: Skill Interpret Graphs 27
Lesson 5-5 Line Graphs 28
Lesson 5-6 Stem-and-Leaf Plots 29
Lesson 5-7 Make an Appropriate Graph 30

Lesson 6-1 Range, Mean, Median, and Mode 31
Lesson 6-2 Choose the Most Appropriate Average 32
Lesson 6-3 Explore Sampling 33
Lesson 6-4 Sampling 34
Lesson 6-5 Problem Solving: Strategy Do an Experiment 35

Lesson 7-1 Divisibility 36
Lesson 7-2 Explore Prime and Composite Numbers 37
Lesson 7-3 Greatest Common Factors and Least Common Multiples 38
Lesson 7-4 Understanding Fractions 39
Lesson 7-5 Simplify Fractions 40
Lesson 7-6 Problem Solving: Skill Extra or Missing Information 41

Lesson 8-1 Compare and Order Fractions 42
Lesson 8-2 Problem Solving: Strategy Make a Table 43
Lesson 8-3 Mixed Numbers 44
Lesson 8-4 Relate Fractions and Decimals 45
Lesson 8-5 Compare and Order Fractions and Mixed Numbers 46

Lesson 9-1 Add and Subtract Fractions with Like Denominators 47
Lesson 9-2 Explore Adding and Subtracting Fractions with Unlike Denominators ... 48
Lesson 9-3 Add Fractions with Unlike Denominators 49
Lesson 9-4 Subtract Fractions with Unlike Denominators 50
Lesson 9-5 Problem Solving: Skill Multistep Problems 51

Lesson 10-1 Add Mixed Numbers 52
Lesson 10-2 Explore Subtracting Mixed Numbers 53
Lesson 10-3 Subtract Mixed Numbers 54
Lesson 10-4 Problem Solving: Strategy Find a Pattern 55
Lesson 10-5 Estimate Sums and Differences 56

Lesson 11-1 Fractions of Whole Numbers 57
Lesson 11-2 Explore Multiplying Fractions 58
Lesson 11-3 Multiply Fractions 59
Lesson 11-4 Multiply Mixed Numbers 60
Lesson 11-5 Problem Solving: Skill Choose the Operation 61

Lesson 12-1 Estimate Products and Quotients 62
Lesson 12-2 Explore Dividing by Fractions 63
Lesson 12-3 Divide Fractions and Mixed Numbers 64
Lesson 12-4 Problem Solving: Strategy Work Backward 65

Lesson 13-1 Time 66
Lesson 13-2 Customary Length 67
Lesson 13-3 Customary Capacity and Weight 68
Lesson 13-4 Problem Solving: Skill Choose an Appropriate Unit 69

Lesson 14-1 Metric Length 70
Lesson 14-2 Metric Capacity and Mass 71
Lesson 14-3 Problem Solving: Strategy Draw a Diagram 72
Lesson 14-4 Explore Conversions Between Systems 73
Lesson 14-5 Temperature 74

Lesson 15-1 Algebra: Order of Operations 75
Lesson 15-2 Algebra: Functions 76
Lesson 15-3 Algebra: Sequences and Functions 77
Lesson 15-4 Algebra: Graph a Function 78
Lesson 15-5 Algebra: Linear and Nonlinear Functions 79
Lesson 15-6 Problem Solving: Skill Describe Relationships 80

Lesson 16-1 Algebra: Explore Addition Equations 81
Lesson 16-2 Algebra: Addition and Subtraction Equations 82
Lesson 16-3 Algebra: Multiplication and Division Equations 83
Lesson 16-4 Algebra: Two-Step Equations 84
Lesson 16-5 Problem Solving: Strategy Write an Equation 85
Lesson 16-6 Algebra: Use Properties of Operations 86
Lesson 16-7 Algebra: Explore Inequalities 87

Lesson 17-1 Algebra: Integers and the Number Line 88
Lesson 17-2 Algebra: Explore Adding Integers 89
Lesson 17-3 Algebra: Add Integers 90
Lesson 17-4 Algebra: Explore Subtracting Integers 91
Lesson 17-5 Algebra: Subtract Integers 92
Lesson 17-6 Problem Solving: Skill Check for Reasonableness 93
Lesson 17-7 Algebra: Multiply and Divide Integers 94

Lesson 18-1 Negative Exponents 95
Lesson 18-2 Scientific Notation 96
Lesson 18-3 Problem Solving: Strategy Alternate Solution Methods 97
Lesson 18-4 Algebra: Rational Numbers 98
Lesson 18-5 Algebra: Operations on Rational Numbers 99
Lesson 18-6 Algebra: Graph Equations in Four Quadrants 100
Lesson 18-7 Algebra: Perfect Squares and Square Roots 101

Lesson 19-1 Basic Geometry Terms 102
Lesson 19-2 Measure and Classify Angles 103
Lesson 19-3 Lines and Angles 104
Lesson 19-4 Triangles 105
Lesson 19-5 Quadrilaterals 106
Lesson 19-6 Problem Solving: Skill Draw a Diagram 107
Lesson 19-7 Circles 108

Lesson 20-1 Congruence and Similarity 109
Lesson 20-2 Transformations 110
Lesson 20-3 Symmetry.. 111
Lesson 20-4 Problem Solving: Strategy Find a Pattern 112
Lesson 20-5 Constructions 113
Lesson 20-6 Explore Tessellations.. 114

Lesson 21-1 Algebra: Perimeter 115
Lesson 21-2 Algebra: Area of Rectangles and Parallelograms 116
Lesson 21-3 Algebra: Area of Triangles and Trapezoids................................... 117
Lesson 21-4 Algebra: Explore Circumference of Circles 118
Lesson 21-5 Algebra: Explore Area of Circles 119
Lesson 21-6 Algebra: Circumference and Area of Circles................................. 120
Lesson 21-7 Problem Solving: Strategy Solve a Simpler Problem 121

Lesson 22-1 3-Dimensional Figures 122
Lesson 22-2 Algebra: Surface Area of Prisms 123
Lesson 22-3 Algebra: Volume of Prisms 124
Lesson 22-4 Problem Solving: Strategy Use Logical Reasoning 125
Lesson 22-5 Algebra: Explore Surface Area of Cylinders 126
Lesson 22-6 Algebra: Explore Volume of Cylinders....................................... 127

Lesson 23-1 Ratios and Equivalent Ratios 128
Lesson 23-2 Rates 129
Lesson 23-3 Better Buy 130
Lesson 23-4 Problem Solving: Skill Check for Reasonableness............................ 131

Lesson 24-1 Algebra: Proportions 132
Lesson 24-2 Problem Solving: Strategy Use a Graph 133
Lesson 24-3 Explore Similar Figures 134
Lesson 24-4 Scale Drawings and Maps... 135

Lesson 25-1 Explore the Meaning of Percent 136
Lesson 25-2 Percents, Fractions, and Decimals.. 137
Lesson 25-3 Algebra: Percent of a Number.. 138
Lesson 25-4 Problem Solving: Skill Choose a Representation 139
Lesson 25-5 Algebra: Find the Percent One Number Is of Another 140

Lesson 26-1 Sales Tax and Discounts... 141
Lesson 26-2 Simple Interest 142
Lesson 26-3 Circle Graphs 143
Lesson 26-4 Problem Solving: Strategy Logical Reasoning 144

Lesson 27-1 Probability... 145
Lesson 27-2 Explore Experimental Probability 146
Lesson 27-3 Make Predictions 147
Lesson 27-4 Mutually Exclusive Events.. 148
Lesson 27-5 Problem Solving: Skill Check for Reasonableness............................ 149

Lesson 28-1 Compound Events 150
Lesson 28-2 Problem Solving: Strategy Make an Organized List.......................... 151
Lesson 28-3 Explore Independent and Dependent Events 152
Lesson 28-4 Independent and Dependent Events 153

Summer Skills Refresher
Number Sense, Concepts, and Operations 157
Measurement.. 159
Geometry and Spatial Sense.. 161
Algebraic Thinking 163
Data Analysis and Probability 165

Name _____

Place Value

P 1-1 PRACTICE

Write each number in words.

1. 34,671,111,800 _____

_____

2. 84,508 _____

3. 12.873 _____

4. 8.0552 _____

5. 0.0065 _____

Write the place and the value of the digit **7** in each number.

6. 0.7 _____

7. 4.00712 _____

8. 2.179 _____

9. 28,467,089 _____

10. 348.92971 _____

Write each number in expanded form.

11. 779,000,400 _____

_____

12. 32,000,000 _____

Write each number in standard form.

13. eleven billion _____

14. 921 million, 780 thousand, 33 _____

15. six and five thousandths _____

16. nine hundred fifty-four ten-thousandths _____

Problem Solving
Solve.

17. A newspaper reporter is covering a gymnastics competition. In his story he wrote that a local gymnast had scored 9.055. Write the score in word form.

_____

18. A 30-second commercial during the broadcast of the 1999 Super Bowl cost one million, six hundred thousand dollars. Write the cost in standard form.

_____

Use with Grade 6, Chapter 1, Lesson 1, pages 2–4.

1

Name_____

Compare and Order

1-2 PRACTICE

Compare. Write all the symbols that would make a true sentence:
<, >, =, ≤, and ≥.

1. 0.62 ◯ 0.618
2. 9.8 ◯ 9.80
3. 1.006 ◯ 1.02

4. 41.3 ◯ 41.03
5. 2.01 ◯ 2.011
6. 1.400 ◯ 1.40

7. 5.079 ◯ 5.08
8. 12.96 ◯ 12.967
9. 15.8 ◯ 15.800

10. 7.98 ◯ 7.89
11. 15 ◯ 15.01
12. 32,174 ◯ 32,740

13. 8,917 ◯ 8,907
14. 11,560,561 ◯ 11.5671
15. 0.01 ◯ 0.001

Graph each set of numbers on a number line.

16. 0.05, 0.61, 0.89

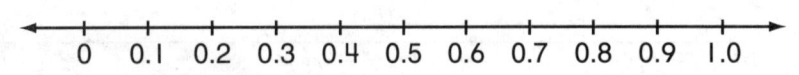

0   0.1   0.2   0.3   0.4   0.5   0.6   0.7   0.8   0.9   1.0

17. 0.26, 0.03, 0.65, 0.59, 0.76

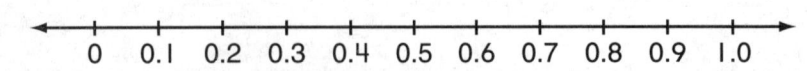

0   0.1   0.2   0.3   0.4   0.5   0.6   0.7   0.8   0.9   1.0

Order from least to greatest.

18. 18,715; 19,175; 18,350 _____

19. 0.5; 0.505; 0.55 _____

20. 654,850; 6,541,850; 651,893 _____

21. 92,018; 92,810; 92,180 _____

22. 4.032; 4.023; 4.203 _____

Problem Solving
Solve.

23. Five divers have entered a competition. Four of the divers have had a turn. The scores are 9.80, 9.75, 9.81, and 9.79. What is the lowest score with 3 digits that the fifth diver can get to win the competition?

24. The times for the first three runners of the 100-yard dash are 9.85 seconds, 9.62 seconds, and 9.60 seconds. What is the winning time? What is the time for the second-place runner?

_____

_____

2

Use with Grade 6, Chapter 1, Lesson 2, pages 6–7.

Add and Subtract Whole Numbers and Decimals

P 1-3 PRACTICE

Add or subtract.

1. 16.982 + 15.19 = _____
2. 322.96 − 19.982 = _____
3. 612.98 − 43.301 = _____
4. 182.91 + 62.29 = _____
5. 98.83 − 62.019 = _____
6. 421.053 + 16.59 = _____
7. 198.6 − 23 = _____
8. 667.1 + 92.54 = _____
9. 83 − 6.98 = _____
10. 4.392 + 76.933 = _____
11. 32.016 − 9.983 = _____
12. 144.53 + 33.801 = _____

13. 87,541
    + 9,689

14.   234.50
    + 187.95

15. 315.93
    − 24.65

16. 12,875
    − 6,459

Find the output.

17. Rule: Subtract 3.75

Input	Output
36.8	
172	
3.809	
5.4	
28.19	

18. Rule: Add 10.4

Input	Output
3.7	
8.25	
82.5	
156	
47.62	

Problem Solving
Solve.

19. An Olympic ski jumper participated in three different jumps. The points for the three jumps were 85.39, 92.60, and 89.07. What was the ski jumper's total score?

20. Andrew is training for the 200-meter dash. Yesterday he ran the dash in 31.25 seconds. He wants to run the dash in 28 seconds. In how much less time must he run in order to reach his goal?

Use with Grade 6, Chapter 1, Lesson 3, pages 8–10.

Name_____

Estimate Sums and Differences

P 1-4 PRACTICE

Round each number to the place indicated.

tens

1. 793 _____
2. 795 _____

hundreds

3. 582 _____
4. 1,104 _____

thousands

5. 8,543 _____
6. 2,264 _____

dollars

7. $91.95 _____
8. $207.48 _____

cents

9. $3.242 _____
10. $0.555 _____

tenths

11. 3.386 _____
12. 4.555 _____

hundredths

13. 0.912 _____
14. 0.3353 _____

thousandths

15. 3.7004 _____
16. 0.31251 _____

Add or subtract. Estimate to check for reasonableness. Show your estimate.

17. $8.02 + $5.19 _____
18. 3.25 − 0.92 _____
19. 432.5 + 12.53 _____
20. 63.4 − 21.07 _____
21. $4.05 + $3.98 _____
22. $1.87 + $5.28 + $9.05 _____
23. 86.57 + 54.38 _____
24. 489.65 − 309.6 _____
25. 7.081 + 5.011 + 4.92 _____
26. $896.53 − $43.42 _____
27. 42.2 + 4.51 _____
28. 83.62 + 9.41 _____
29. 10.84 − 1.79 _____
30. 36.15 − 6.63 _____

Problem Solving

Solve.

31. Carla buys two magazines that cost $4.99 each and a notebook that costs $2.15. How much do the items cost altogether? Explain how you estimated to check if your answer is reasonable.

_____

_____

4

Use with Grade 6, Chapter 1, Lesson 4, pages 12–14.

Name_____

Problem Solving: Skill
Using the Four-Step Process

P 1-5 PRACTICE

Solve. Use data from the table to solve problems 1–2.

Women's Olympic 50-Meter Freestyle		
Year	Gold Medal Winner	Time (in seconds)
1992	Yang Wenyi	24.76
1996	Amy Van Dyken	24.87
2000	Inge de Bruijn	24.32

1. How much less than Amy Van Dyken's time was Yang Wenyi's time?
   _____

2. Which gold medal winner in the table had the fastest time?
   _____

Use data from the table below to solve problems 3–6.

Women's Olympic 100-Meter Butterfly		
Year	Gold Medal Winner	Time (in seconds)
1988	Kristin Otto	59.00
1992	Qian Hong	58.62
1996	Amy Van Dyken	59.13
2000	Inge de Bruijn	56.61

3. How much less than Qian Hong's time was Inge de Bruijn's time?
   _____

4. Which gold medal winner had the slowest time?
   _____

5. Which gold medal winner had the fastest time?

   A Kristin Otto

   B Qian Hong

   C Amy Van Dyken

   D Inge de Bruijn

6. Which number sentence shows the difference between the fastest swimmer and the slowest swimmer?

   F 59.13 − 58.62 = 0.51

   G 59.13 − 56.61 = 2.52

   H 59.00 − 56.61 = 2.39

   J 58.62 − 56.61 = 2.01

Use with Grade 6, Chapter 1, Lesson 5, pages 16–17.

Name _____

Explore Addition and Subtraction Expressions

P 2-1 PRACTICE

Evaluate each expression for the value given. Use models if you wish.

1. $x + 22$, for $x = 10$ _____

2. $132 + n$, for $n = 56$ _____

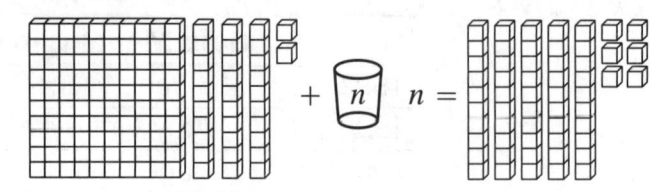

3. $x - 15$, for $x = 50$ _____

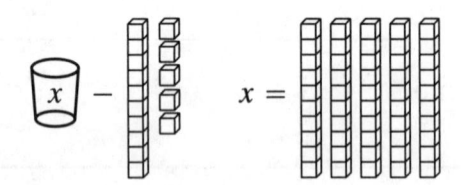

4. $127 - c$, for $c = 60$ _____

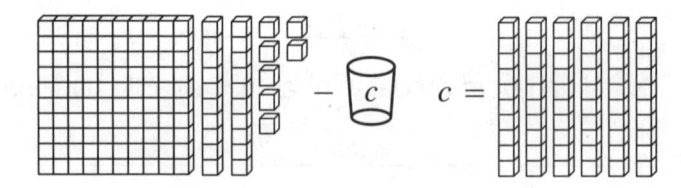

5. $100 - n$, for $n = 18$ _____

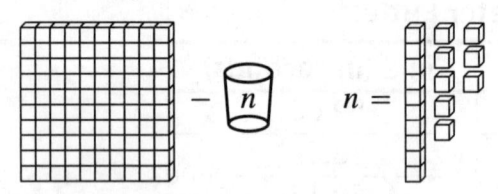

6. $t + 112$, for $t = 25$ _____

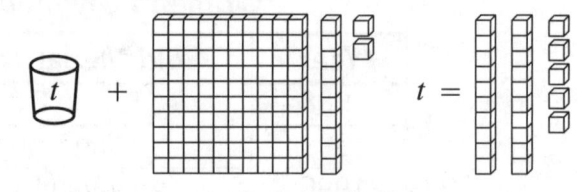

7. $a + 45$, for $a = 16$ _____

8. $1,999 + x$, for $x = 25$ _____

9. $x - 437$, for $x = 500$ _____

10. $12 - n$, for $n = 3.05$ _____

11. $7.8 + v$, for $v = 10$ _____

12. $4.1 - c$, for $c = 4.1$ _____

13. $16 - t$, for $t = 4.25$ _____

14. $x + 2.378$, for $x = 0.02$ _____

Problem Solving
Solve.

15. Rosita was born on her brother's birthday. Her brother is 6 years older than Rosita. If Rosita's brother is x years old, what expression represents Rosita's age?

16. Todd counted t items in his mother's shopping cart. She bought seven more items. Write an expression to show the total number of items.

_____

_____

6

Use with Grade 6, Chapter 2, Lesson 1, pages 22–23.

Name _____

Problem Solving: Strategy
Write an Equation

2-2 PRACTICE

Solve.

1. Matt bought a tennis racket that usually costs $73.95. He had a coupon for a discount of *d* dollars. The net price of the racket with the discount was *c* dollars. Write an equation that represents the relationship between the net price and the discount.

   _____

2. Ms. Gonzaga ordered a bookcase that cost $89.45. The delivery fee was *f* dollars. The cost with the delivery fee was *t* dollars. Write an equation that represents the relationship between the delivery fee and the cost with the delivery fee.

   _____

3. Use the equation you wrote for problem 1 to find the net price if the discount was $7.50.

   _____

4. Use the equation you wrote for problem 2 to find the total cost if the delivery fee was $29.95.

   _____

Mixed Strategy Review

Solve. Use any strategy.

5. **Number Sense** Brooke is making a necklace in which the first, fifth, ninth, and thirteenth beads are blue and the rest of the first 15 beads are not blue. If the necklace continues this pattern and has 50 beads in all, how many of them will be blue?

   _____

   Strategy: _____

6. **Career** A salesman spends $89 per night for 5 nights at a hotel, $219.49 for transportation, and $137.71 for food. What are his total travel expenses?

   _____

   Strategy: _____

7. Pablo had $5.50 left after spending $7.50 for a set of colored pencils and $3.25 for a drawing pad. How much money did Pablo have to start with?

   _____

   Strategy: _____

8. **Create a problem** that you could solve using an equation. Write the equation. Share your work with others.

   _____

   _____

Use with Grade 6, Chapter 2, Lesson 2, pages 24–25.

Name _____

Properties of Addition

P 2-3 PRACTICE

Identify the property of addition used.

1. 12 + 9 = 9 + 12 = 21 _____
2. 91 + (59 + 37) = (91 + 59) + 37 = 150 + 37 = 187 _____
3. 47 + 0 = 47 _____
4. (1.05 + 2.4) + 1.6 = 1.05 + (2.4 + 1.6) = 1.05 + 4 = 5.05 _____

Complete. Name the property of addition used.

5. 12 + 18 = 18 + ☐

6. (☐ + 0) + 5.8 = 7.21 + 5.8

7. (8 + 9) + 11 = ☐ + (9 + 11)

8. 23.05 + ☐ = 17 + 23.05

9. (1.5 + ☐) + 1.6 = 1.5 + (2.31 + 1.6)

10. 7.5 + 8.3 = ☐ + 7.5 + 8.3

11. 42.8 + (7.58 + ☐) = 42.8 + (5.61 + 7.58)

12. 15.8 + 23.7 = ☐ + 15.8 + 23.7

Use mental math or paper and pencil to find each sum.

13. 36 + 9 + 14 = _____
14. 42.7 + 0 = _____
15. 46 + 0 + 24 = _____
16. 35.87 + 0 = _____
17. 14 + 33 + 26 + 17 = _____
18. 5.9 + 0.7 + 1.1 = _____
19. 17.5 + 14.2 + 5.8 = _____
20. 15.8 + 7.9 + 4.2 = _____
21. 3.6 + 7.2 + 1.4 + 4.8 = _____
22. 33.7 + 0 + 42.3 = _____
23. 43 + 65 + 37 + 55 = _____
24. 11.4 + 7.8 + 6.6 + 3.2 = _____
25. 10.08 + 0 + 4.02 = _____
26. 13 + 56 + 87 + 44 = _____

Problem Solving

Solve.

27. In four football games, Fred gained 125 yards, 100 yards, 75 yards, and 115 yards. What is the total gain for the four games?

28. The times for three runners in a relay race are 63 seconds, 57 seconds, and 30 seconds. What is the difference between the fastest time and the slowest time?

_____

Use with Grade 6, Chapter 2, Lesson 3, pages 26–28.

Name_____

Choose a Computation Method

P 2-4 PRACTICE

Add or subtract. Tell which method you used and explain why you chose that method.

_____

_____

1. 929 + 300 = _____

2. 84.1 − 25.9 = _____

3. 4,592 + 2,002 = _____

4. 6,315 − 500 = _____

5. $45.99 − $4.09 = _____

6. $5.12 + $1.98 = _____

7. 73.82 − 19.116 = _____

8. 47.2 + 1.1 = _____

9. 52 − 9.5 = _____

10. $28.00 − $6.23 = _____

11. 52 + 306 + 48 = _____

12. 5,792 + 82.7 = _____

13. 768.3 − 15.2 = _____

14. $25.95 − $15.01 = _____

15. $7.48 + $4.39 = _____

Problem Solving
Solve.

16. Francesca bought a skateboard for $64.99. The sales tax was $3.89. What was the total, including sales tax?

17. Emilio buys a concert ticket for $15.75, including tax. How much change should he receive from $20?

Use with Grade 6, Chapter 2, Lesson 4, pages 30–31.

9

Name _____

Multiplication Patterns

P 3-1 PRACTICE

Multiply. Use mental math.

1. 6 × 100 = _____
2. 8 × 300 = _____
3. 20 × 50 = _____
4. 4 × 600 = _____
5. 1,000 × 23 = _____
6. 900 × 20 = _____
7. 800 × 60 = _____
8. 12 × 60 = _____
9. 12 × 5,000 = _____
10. 500 × 90 = _____
11. 11 × 300 = _____
12. 70 × 600 = _____
13. 60 × 50 = _____
14. 80 × 200 = _____
15. 90 × 70 = _____
16. 100 × 90 = _____
17. 600 × 12 = _____
18. 40 × 90 = _____
19. 50 × 700 = _____
20. 70 × 300 = _____
21. 40 × 80 = _____
22. 70 × 110 = _____

Find each missing number.

23. 7 × _____ = 4,900
24. 6 × _____ = 540
25. 15 × _____ = 15,000
26. 90 × _____ = 4,500
27. 40 × _____ = 2,400
28. 8 × _____ = 2,400

Problem Solving
Solve.

29. Dina had a summer job at a restaurant for 70 days. She worked 6 hours a day. How many hours did she work that summer?

30. Karla travels 7 miles on her bike each day to deliver *The Daily News*. If she completes her route every day of September, how many miles will she travel that month?

10

Use with Grade 6, Chapter 3, Lesson 1, pages 46–47.

Name_____

Multiply Whole Numbers

P 3-2 PRACTICE

Multiply.

1. 142 × 6 = _____

2. 7 × 407 = _____

3. 8 × $396 = _____

4. 2,197 × 29 = _____

5. 67,418 × 64 = _____

6. 69 × 46 = _____

7. 57 × $37 = _____

8. 656 × 23 = _____

9. 390 × 48 = _____

10. 3,507 × 54 = _____

11. 378 × 76 = _____

12. 4,760 × 93 = _____

13. 73 × $547 = _____

14. 406 × 326 = _____

15. 327 × 544 = _____

16. 215 × 58 = _____

17. 19 × $739 = _____

18. 862 × 12 = _____

19. 4,925 × 6 = _____

20. 167 × 329 = _____

21. 3,819 × 600 = _____

22. 21,934 × 413 = _____

23. 85 × 63 = _____

24. 472 × $543 = _____

25. 45 × 306 = _____

26. 208 × 415 = _____

Problem Solving
Solve.

27. Karen's family bought 14 rolls of film for their vacation. There are 24 pictures on each roll of film. If they used all except 3 rolls, how many pictures did they take?

_____

28. Jill took a bike tour of France. She rode her bike 12 miles each day. After 27 days, how many miles had Jill ridden?

_____

29. A bus tour will cover 365 miles each day for 11 days. How many miles will the bus travel?

_____

30. One travel agency is advertising round-trip tickets for $238. They book 56 tickets in one week. How much do these tickets cost altogether?

_____

Use with Grade 6, Chapter 3, Lesson 2, pages 48–49.

11

Name _____

Exponents

P 3-3 PRACTICE

Write using an exponent.

1. $8 \times 8 \times 8 \times 8 \times 8$ _____
2. $3 \times 3 \times 3 \times 3 \times 3 \times 3$ _____
3. 7×7 _____
4. $1 \times 1 \times 1 \times 1 \times 1 \times 1 \times 1 \times 1$ _____
5. $4 \times 4 \times 4 \times 4 \times 4$ _____
6. 3×3 _____
7. $6 \times 6 \times 6$ _____
8. $7 \times 7 \times 7 \times 7 \times 7 \times 7 \times 7 \times 7 \times 7$ _____
9. $5 \times 5 \times 5 \times 5$ _____
10. $10 \times 10 \times 10$ _____
11. $46 \times 46 \times 46$ _____
12. 5×5 _____
13. $1 \times 1 \times 1$ _____
14. 12×12 _____

Evaluate.

15. 3^5 _____
16. 6^3 _____
17. 7^2 _____
18. 10^4 _____
19. 5^4 _____
20. 4^5 _____
21. 7^3 _____
22. 6^5 _____
23. 8^4 _____
24. 10^7 _____
25. 9^2 _____
26. 3^4 _____

Problem Solving
Solve.

27. What is the cube of 3?

28. What is the square of 25?

29. What is 5 cubed?

30. What is 10 to the sixth power?

31. Suppose a cell splits into two cells every hour. Starting with 1 cell, how many of these cells will there be after 12 hours? Express your answer using an exponent.

32. The memory of a computer is measured in a unit called a byte. The letter K represents 2^{10} bytes. Write this number without an exponent.

Use with Grade 6, Chapter 3, Lesson 3, pages 50–51.

Name _____

Estimate Products

P 3-4 PRACTICE

Multiply. Estimate to check if your answer is reasonable. Show your estimate.

1. 542
   × 3.8

2. $8.21
   × 14

3. 72.6
   × 38

4. 17.4
   × 7.3

5. $28.62
   ×   143

6. 12.649
   × 38.2

7. $692.38
   ×   54

8. 104.05
   × 63.2

Estimate each product.

9. 1.4 × 7.7 _____

10. 3.4 × 87.3 _____

11. 46.9 × 1.8 _____

12. 89 × $8.06 _____

13. 4.7 × 96.2 _____

14. 37.21 × 49.9 _____

15. 198 × $22.03 _____

16. 39.26 × 1.98 _____

17. 18.8 × 4.3 _____

18. 2.18 × 24.19 _____

19. 41 × $18.75 _____

20. 0.91 × 15.8 _____

21. 63 + 59 + 64 + 57 + 61 + 62 = _____

22. 215 + 196 + 203 + 212 = _____

Problem Solving
Solve.

23. Harry earns $4.75 per hour mowing lawns. About how much will he earn in 3.5 hours?

24. Jared saves $6.25 each week. About how much will he save in one year?

Use with Grade 6, Chapter 3, Lesson 4, pages 52–54.

Name _____

Problem Solving: Skill
Estimate or Exact Answer

P 3-5
PRACTICE

Solve. Did you give an estimate or exact answer? Explain.

1. It costs Matt a little more that $4 a day to feed his dog. How much does it cost him to feed his dog for a year?

   _____

2. In the past year, a grocery store deposited about 6 million pennies, 3 million nickels, 4 million dimes, and 2 million quarters in the bank. What is the total value of the deposit?

   _____

3. A bank puts 3,000 quarters in each bag. How much are 15 bags of quarters worth?

   _____

4. A vault contains $3,000 worth of nickels. How many nickels are in the vault?

   _____

5. When at rest, your heart probably beats about 70 times per minute. At that rate, how many times does it beat in an hour?

   _____

6. Ann bought two shirts for $28.95 and a skirt for $33.95. The sales tax was $3.71. How much did she pay altogether?

   _____

Choose the correct answer.

It costs $0.38 to produce and mail a newsletter. Each week, 475,000 newsletters are mailed to subscribers.

7. What is the cost of producing and sending the newsletters for a three month period? _____

8. Which of the following statements is true?

   A The cost of producing and sending newsletters for one month is about $2,000,000.

   B More than 12,000,000 newsletters are produced and mailed in a three-month period.

   C The cost for two months is about $1,444,000.

   D About 160,000 newsletters are produced and mailed each month.

9. If the numbers in a problem appear to be rounded, you can

   F find an exact answer.

   G estimate the answer.

   H ignore the numbers in the problem.

   J check your answer.

14

Use with Grade 6, Chapter 3, Lesson 5, pages 56–57.

Name _____

Multiply Decimals

P 3-6 PRACTICE

Multiply. Estimate to check if your answer is reasonable.

1. 4.6 × 263	2. 0.92 × 87	3. 0.024 × 6	4. 0.38 × 67	5. 21 × 3

6. 42 ×0.056	7. 0.092 × 5.5	8. 2.031 × 2.1	9. 5.3 × 5	10. 0.53 × 5

11. 0.053 × 5 = _____

12. 1.9 × 9 = _____

13. 2.065 × 12 = _____

14. 25 × 35.15 = _____

15. 5.6 × 3.1 = _____

16. 6.108 × 3.5 = _____

17. 7.13 × 8 = _____

18. 0.26 × 1.2 = _____

19. 6.4 × 5.7 = _____

20. 6.3 × 1.075 = _____

Compare. Write >, <, or =.

21. 0.5 × 0.5 ◯ 0.3 × 0.9

22. 0.6 × 0.8 ◯ 0.7 × 0.7

23. 0.8 × 0.8 ◯ 0.9 × 0.7

24. 0.9 × 0.9 ◯ 0.9

25. 0.5 × 0.6 ◯ 0.3

26. 1.0 ◯ 0.9 × 0.12

Problem Solving
Solve.

27. A ream of paper consists of 500 sheets. The thickness of one sheet of paper is 0.01 cm. Calculate the thickness of a ream of paper.

28. Human hair grows at a rate of about 0.5 inch per month. At this rate, how long will a person's hair grow in 7 months?

29. A dollar bill is about 0.0043 inch thick. How thick would a stack of 25 one-dollar bills be?

30. A travel poster is 0.0048 centimeter thick. How thick is a stack of 6 travel posters?

Use with Grade 6, Chapter 3, Lesson 6, pages 58–60.

Name _____

Properties and Mental Math • Algebra

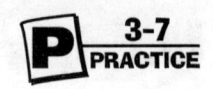

P PRACTICE 3-7

Multiply. Use mental math. Name the property of multiplication you used.

1. $5 \times 102 =$ _____

2. $8 \times 88 =$ _____

3. $5 \times 24 =$ _____

4. $6 \times \$44 =$ _____

5. $8 \times 5.2 =$ _____

6. $8 \times 51 =$ _____

7. $3.3 \times 9 =$ _____

8. $0.82 \times 8 =$ _____

9. $4.5 \times 7 =$ _____

10. $717 \times 0 \times 8 =$ _____

11. $483 \times 1 \times 10 =$ _____

12. $5 \times 755 \times 2 =$ _____

13. $5 \times 9 \times 5 =$ _____

14. $0.7 \times 8 \times 4 =$ _____

15. $8.2 \times 8 =$ _____

16. $8 \times 0 \times 9 \times 34 =$ _____

Use mental math, paper and pencil, or a calculator to find each product.

17. $56 \times 53 =$ _____

18. $8.2 \times 41 =$ _____

19. $83 \times 90 =$ _____

20. $28 \times 6.2 =$ _____

21. $8 \times \$9.84 =$ _____

22. $7 \times \$8.65 =$ _____

23. $81 \times 46 =$ _____

24. $240 \times 40 =$ _____

25. $18 \times 18 =$ _____

Complete. Name the property of multiplication used.

26. $(185 \times 6) \times$ _____ $= 185 \times (6 \times 2)$

27. $9 \times (60 + 7) = ($_____$\times 60) + (9 \times 7)$

28. $124 \times$ _____ $= 14 \times 124$

29. $3.41 \times$ _____ $= 3.41$

Name _____

Estimate Quotients

P 4-1 PRACTICE

Divide. Use mental math.

1. 2,100 ÷ 7 = _____
2. 2,100 ÷ 70 = _____
3. 2,100 ÷ 700 = _____
4. 540 ÷ 90 = _____
5. 270 ÷ 9 = _____
6. 3,500 ÷ 50 = _____
7. 1,600 ÷ 20 = _____
8. 3,500 ÷ 700 = _____
9. 6,000 ÷ 300 = _____
10. 8,400 ÷ 120 = _____
11. 16,000 ÷ 400 = _____
12. 45,000 ÷ 5,000 = _____
13. 6,000 ÷ 50 = _____
14. 1,800 ÷ 30 = _____
15. 4,200 ÷ 60 = _____
16. 150,000 ÷ 3 = _____
17. 480,000 ÷ 60,000 = _____
18. 18,000 ÷ 300 = _____

Estimate each quotient.

19. 717 ÷ 9 _____
20. 349 ÷ 7.4 _____
21. 638 ÷ 8.2 _____

22. 463 ÷ 90 _____
23. 249 ÷ 81 _____
24. 2,322 ÷ 70 _____

25. 12.398 ÷ 4.1 _____
26. 162.58 ÷ 4.4 _____
27. 534.82 ÷ 9.1 _____

28. 56.784 ÷ 7.9 _____
29. 481.4 ÷ 64 _____
30. 34.594 ÷ 4.6 _____

31. 57.36 ÷ 9.32 _____
32. 468.65 ÷ 8.43 _____
33. 409.4 ÷ 48.4 _____

34. 363.68 ÷ 3.6 _____
35. 311.24 ÷ 5.2 _____
36. 364.56 ÷ 69.3 _____

Problem Solving
Solve.

37. Jane makes 20 equal payments to buy a CD player that sells for $180. How much is each payment?

38. Justine makes 30 equal payments to buy a car that sells for $15,000. How much is each payment?

_____

_____

Use with Grade 6, Chapter 4, Lesson 1, pages 70–73.

17

Name _____

Divide Whole Numbers

P 4-2 PRACTICE

Divide. Estimate to check that your answer is reasonable.

1. 54)4,106
2. 7)862
3. 29)1,775

4. 3)195
5. 98)62,000
6. 6)918

7. 66)2,310
8. 53)3,803
9. 45)4,369

10. 145)1,701
11. 6)45,380
12. 654)7,102

Use mental math, paper and pencil, or a calculator to divide.
Show the remainder as a whole number. Check your answer.

13. 4,875 ÷ 82 = _____
14. 2,602 ÷ 37 = _____

15. 2,148 ÷ 62 = _____
16. 8,932 ÷ 451 = _____

17. 3,494 ÷ 349 = _____
18. 9,456 ÷ 295 = _____

19. 27,568 ÷ 9 = _____
20. 5,688 ÷ 34 = _____

Problem Solving
Solve.

21. A family of 4 spent $62.40 for tickets to a soccer game. All of the tickets were the same price. What was the cost of each ticket?

22. A ticket seller collected $1,035 for selling tickets. Each ticket costs $15. How many tickets did she sell?

Use with Grade 6, Chapter 4, Lesson 2, pages 74–77.

Name _____

Divide Decimals by Whole Numbers

P 4-3 PRACTICE

Divide. Round to the nearest tenth.

1. 1.84 ÷ 6 = _____
2. 3.2 ÷ 16 = _____
3. 5.13 ÷ 17 = _____

4. 21.6 ÷ 9 = _____
5. 123 ÷ 15 = _____
6. 108 ÷ 5 = _____

7. 17.5 ÷ 50 = _____
8. 120.6 ÷ 224 = _____
9. 11.4 ÷ 18 = _____

10. 889 ÷ 34 = _____
11. 31.6 ÷ 5 = _____
12. 4.15 ÷ 5 = _____

Divide. Round to the nearest hundredth. Estimate to check if your answer is reasonable.

13. 7.8 ÷ 10 = _____
14. 89.1 ÷ 400 = _____

15. 46.3 ÷ 19 = _____
16. 0.6 ÷ 20 = _____

17. 23.4 ÷ 20 = _____
18. 5 ÷ 18 = _____

19. 26.14 ÷ 4 = _____
20. 7.3 ÷ 5 = _____

21. 0.12 ÷ 8 = _____
22. 3.39 ÷ 6 = _____

23. 1.45 ÷ 20 = _____
24. 12.6 ÷ 29 = _____

Divide using mental math, paper and pencil, or a calculator.

25. 8)13.5
26. 23)4.6
27. 22)147.4

28. 38)88.92
29. 9)487.22
30. 17)5.1

31. 6)18.6
32. 2)99.67
33. 34)274.55

Problem Solving
Solve.

34. A package of 25 pencils costs $5.75. How much does each pencil cost?

35. A car traveled 234.3 miles on 11 gallons of gas. How many miles per gallon did the car average?

_____ _____

Use with Grade 6, Chapter 4, Lesson 3, pages 78–81.

Name_____

Explore Multiplication and Division Expressions • Algebra

4-4 PRACTICE

Write an expression for each problem.

1. Carol bought a glass of juice. Her friend bought twice as much juice. Let j be the amount of juice Carol bought, in ounces. Write an expression that tells how much juice her friend bought, in ounces.

   _____

2. Carol paid 8 times as much as her friend did for her ticket. Let t be the cost, in dollars, of her friend's ticket. Write an expression for the cost, in dollars, of Carol's ticket.

   _____

3. Sarah runs 6 times a week. How many times does she run in x weeks?

   _____

4. A child's stride is k inches. Her father's stride is 3 times as long. Write an expression for the father's stride, in inches.

   _____

5. A good standing broad jump might be h inches. Write an expression for the number of feet this would equal.

   _____

6. A car traveled 350 miles in h hours. How many miles per hour did it travel?

   _____

Evaluate each expression.

7. $3b$; $b = 3$ _____

8. $4c$; $c = 7$ _____

9. $30c$; $c = 6.5$ _____

10. $d \div 5$; $d = 11$ _____

11. $5a$; $a = 5$ _____

12. $7.85g$; $g = 4$ _____

13. $1.58b$; $b = 12$ _____

14. $130.41 \div b$; $b = 3$ _____

15. $14x$; $x = 7.8$ _____

16. $4.5y$; $y = 128$ _____

17. $1.6x$; $x = 6.4$ _____

18. $122.4 \div y$; $y = 68$ _____

19. $458.64 \div z$; $z = 12$ _____

20. $2.5w$; $w = 72.54$ _____

21. $152.1 \div n$; $n = 6$ _____

22. $t \div 6$; $t = 8.52$ _____

Name _____

Problem Solving: Strategy
Guess and Check

4-5 PRACTICE

Solve using the guess-and-check method.

1. Mary has two main flower beds, one of which is 7 feet longer than the other. She also has a small flower bed that is half the length of her largest bed. If the total length of the flower beds is 48 feet, how long is each of the beds?

   _____

2. Judy is planning her garden. She wants 3 times as many tomato plants as eggplants. She wants twice as many peppers as tomatoes. If she has room for 60 plants in her garden, how many of each kind of plant should she order?

   _____

3. **Career** A truck driver must drive 558 miles in a day to deliver fresh produce to a supermarket. If she drives twice as many miles in the morning as in the afternoon, how many miles does she drive in the afternoon?

   _____

4. **Health** Natalie is training to ride in a bicycle race. She rides 364 miles in 3 days. On the third day, she rides the number of miles she rides the first two days. How many miles does she ride on the third day?

   _____

Mixed Strategy Review
Solve. Use any strategy.

5. **Time** Mark did his English homework in 45 minutes and his math homework in 30 minutes. It took him another 20 minutes to study for a spelling test. If he started his homework at 4:00 in the afternoon, what time did he finish?

   _____

   Strategy: _____

6. Leticia has 8 rows of flowers in her garden. She puts one marigold plant on each end of the first row. On each end of the second row, there are 2 marigold plants. On each end of the third row, there are 3 marigold plants, and so on. How many marigold plants does she need to create her design?

   _____

   Strategy: _____

7. **Mental Math** The sum of two decimals between 0 and 1 is 1. One decimal is 4 times as great as the other decimal. What are the two decimals?

   _____

   Strategy: _____

8. **Create a problem** for which you could use the guess-and-check method to solve. Share it with others.

   _____

   _____

Use with Grade 6, Chapter 4, Lesson 5, pages 84–85.

21

Name _____

Multiply and Divide by Powers of Ten

P PRACTICE 4-6

Complete each pattern.

1. 7.26 × 1 = 7.26
   7.26 × 10 = _____
   7.26 × 100 = _____
   7.26 × 1,000 = _____
   7.26 × 10,000 = _____

2. 0.15 × 1 = 0.15
   0.15 × 10 = _____
   0.15 × 100 = _____
   0.15 × 1,000 = _____
   0.15 × 10,000 = _____

3. 53.8 ÷ 1 = 53.8
   53.8 ÷ 10 = _____
   53.8 ÷ 100 = _____
   53.8 ÷ 1,000 = _____
   53.8 ÷ 10,000 = _____

4. 1,080 ÷ 1 = 1,080
   1,080 ÷ 10 = _____
   1,080 ÷ 100 = _____
   1,080 ÷ 1,000 = _____
   1,080 ÷ 10,000 = _____

Find each product or quotient. Use mental math.

5. 1,000 × 0.12 = _____

6. 0.7 × 10 = _____

7. 0.09 × 100 = _____

8. 0.4 ÷ 1,000 = _____

9. 5.02 ÷ 10 = _____

10. 16.5 ÷ 100 = _____

11. 5.2 × 10 = _____

12. 0.08 × 1,000 = _____

13. 100 × 0.05 = _____

14. 2,367 ÷ 10,000 = _____

15. 45.28 ÷ 10 = _____

16. 0.9 ÷ 1,000 = _____

17. 18.03 × 1,000 = _____

18. 6.1 × 100 = _____

19. 4.7 ÷ 10 = _____

20. 0.07 ÷ 100 = _____

21. 0.32 × 10 = _____

22. 0.03 ÷ 100 = _____

23. 2.6 ÷ 1 = _____

24. 12.6 × 1,000 = _____

25. 0.8 × 1,000 = _____

26. 1,000 × 6.7 = _____

27. 100 × 0.15 = _____

28. 23.5 ÷ 10 = _____

Name _____

Divide Decimals by Decimals

4-7 PRACTICE

Divide. Round to the nearest tenth, if necessary.

1. 3.4)12.92   2. 0.8)26.08   3. 0.67)3.643   4. 0.03)0.294

5. 82.65 ÷ 9.5 = _____   6. 0.476 ÷ 0.6 = _____   7. 34.28 ÷ 0.09 = _____

8. 7.221 ÷ 0.08 = _____   9. 224 ÷ 0.7 = _____   10. 5.1 ÷ 0.003 = _____

11. 0.07)0.868   12. 0.046)3.0084   13. 2.5)8.79   14. 1.3)99

Divide. Round to the nearest hundredth or cent, if necessary.

15. 1.44 ÷ 0.45 = _____   16. 0.3904 ÷ 0.061 = _____

17. 0.5318 ÷ 0.49 = _____   18. 42 ÷ 0.06 = _____

19. 12 ÷ 0.005 = _____   20. 32.2 ÷ 0.46 = _____

21. 63.96 ÷ 7.8 = _____   22. 242 ÷ 0.55 = _____

23. $8.45 ÷ 1.2 = _____   24. 134.4 ÷ 5.1 = _____

25. 41.07 ÷ 0.5 = _____   26. $36.12 ÷ 3.5 = _____

27. $9.99 ÷ 7.3 = _____   28. 18 ÷ 6.75 = _____

Problem Solving
Solve.

29. One type of motor-driven camera can take a picture every 0.06 second. While taking some action pictures, a photographer let the camera run for 3.6 seconds. How many pictures did the camera take?

_____

Use with Grade 6, Chapter 4, Lesson 7, pages 88–91.

Name_____

Collect, Organize, and Display Data

P 5-1 PRACTICE

Use the line plot to answer exercises 1–3.

1. What information is displayed?

   _____

   _____

2. How many students spent time doing homework last night?

   _____

3. How many students spent at least a half hour on homework?

   _____

Time Spent Doing Homework Last Night (minutes)

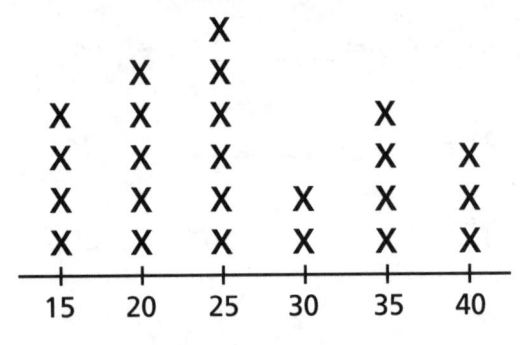

```
              X
           X  X
     X  X  X        X
     X  X  X        X  X
     X  X  X  X  X  X
     X  X  X  X  X  X
    ─┼──┼──┼──┼──┼──┼─
    15 20 25 30 35 40
```

Each X stands for 1 student.

Problem Solving

Solve.

Two bookstores recorded data for the books they sold in the month of June. Use the pictographs to answer exercises 4–9.

Book Nook June Sales

Kind of Books	Books Sold
Adventure	📖📖📖📖
Mystery	📖📖📖
Fantasy	📖📖📖📖📖
Nonfiction	📖
Other	📖📖

Each 📖 stands for 100 books.

Community Bookstore June Sales

Kind of Books	Books Sold
Mystery	📖📖📖📖📖
Fantasy	📖📖📖📖
Adventure	📖📖📖📖📖📖📖
Jokes/Riddles	📖
Other	📖📖📖

Each 📖 stands for 50 books.

4. Which store sold nonfiction books? How many nonfiction books did it sell?

   _____

5. Which store sold more mystery books? How many more mystery books did it sell?

   _____

6. Which kind of book was the best seller for the Community Bookstore?

   _____

7. Which kind of book did the stores sell about the same number of?

   _____

8. About how many books did the Book Nook sell in June?

   _____

9. About how many books did the Community Bookstore sell in June?

   _____

24

Use with Grade 6, Chapter 5, Lesson 1, pages 106–109.

Name _____

Bar Graphs

5-2 PRACTICE

Every week, José tries to improve his swimming times. This double-bar graph shows his best times for two different weeks and three different events.

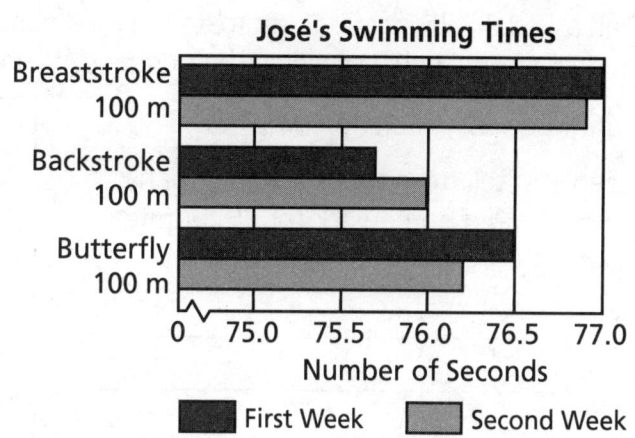

José's Swimming Times

Breaststroke 100 m
Backstroke 100 m
Butterfly 100 m

0 75.0 75.5 76.0 76.5 77.0
Number of Seconds

■ First Week   ▨ Second Week

Use data from the graph for problems 1–3.

1. In how many seconds did José swim the breaststroke the first week? the second week?

   _____

2. For which events did he improve his time the second week? Which showed the greatest improvement?

   _____

3. Approximately what was his best time for the backstroke?

   _____

Tina surveyed sixth graders at her school about their favorite school sport. The table shows the results of her survey. Use data from the table for exercises 4–6.

Favorite School Sport		
Sport	Ms. Kwan's Class	Mr. Brooke's Class
Soccer	12	14
Basketball	7	5
Track	10	11

4. Make a double bar graph to represent the data.

5. How many students chose basketball as their favorite sport?

   _____

6. Which sport was the most popular in Mr. Brooke's class? In Ms. Kwan's class?

   _____

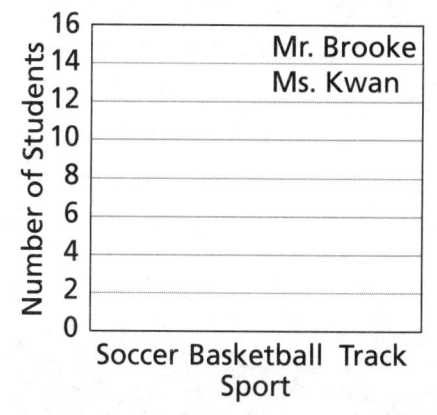

Favorite School Sport

Number of Students: 0, 2, 4, 6, 8, 10, 12, 14, 16

Mr. Brooke
Ms. Kwan

Soccer Basketball Track
Sport

Use with Grade 6, Chapter 5, Lesson 2, pages 110–113.

25

Histograms

5-3 PRACTICE

Nathan asked 24 classmates to estimate the total number of minutes they spent doing homework last night. The frequency table shows their responses.

Use the frequency table to complete.

1. Can you tell from the table how many students did homework for 40 minutes? Explain.

   _____
   _____
   _____
   _____
   _____

| Minutes Spent Doing Homework ||
Number of Minutes	Frequency
0–9	1
10–19	1
20–29	2
30–39	6
40–49	8
50–59	3
60–69	2
70–79	1

2. How many students did homework for at least 30 minutes? _____

3. Make a histogram for the data. Use the intervals in the table.

4. Make a relative frequency table for the data. Round the relative frequencies to the nearest hundredth.

| Minutes Spent Doing Homework ||
Number of Minutes	Relative Frequency

5. Make a histogram to represent the data in the relative frequency table you made.

Name _____

Problem Solving: Skill
Interpret Graphs

5-4 PRACTICE

The graphs show the number of hours that students in Grades 5–8 used the Internet. Use data from the graphs for problems 1–6.

Graph A 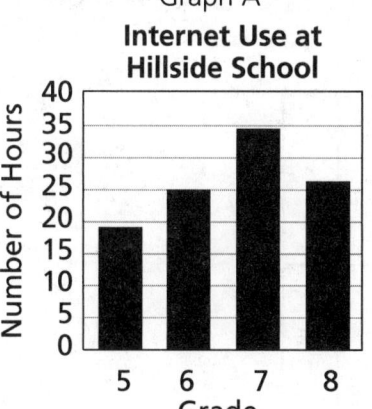 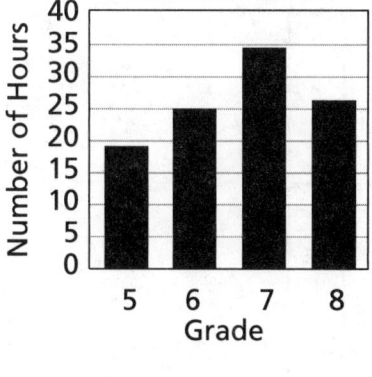
Internet Use at Hillside School

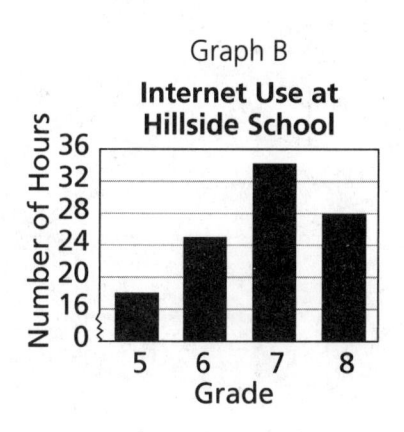

1. What are the intervals for each graph?
   _____

2. Which grade spent the most time on the Internet?
   _____

3. Which graph makes it appear that seventh graders spent about 4 times as many hours on the Internet as fifth graders spent?
   _____

Graph B
Internet Use at Hillside School

4. Which graph makes the differences in internet use appear greater than they are?
   _____

5. Which graph has a break in the vertical scale?
   _____

6. Which graph best represents the data?
   _____

Use the graph at the right for problem 7.

7. Which of the following statements is true?

   A Almost twice as many students prefer cola to lemon.

   B The difference between the number of students who prefer cola and who prefer diet cola is about 500.

   C Fewer than 600 students are represented in the graph.

   D The total number of students who prefer diet cola and lemon is less than the total number who prefer cola.

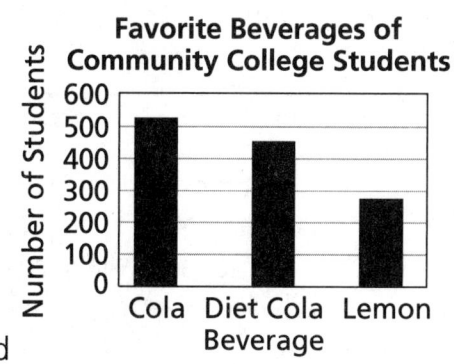

Favorite Beverages of Community College Students

Use with Grade 6, Chapter 5, Lesson 4, pages 118–119.

27

Name_____

Line Graphs

P 5-5 PRACTICE

Sue and Brett each had a lemonade stand. Sue sold pink lemonade and Brett sold regular lemonade. They each charged $0.50 a glass. This double-line graph compares their sales.

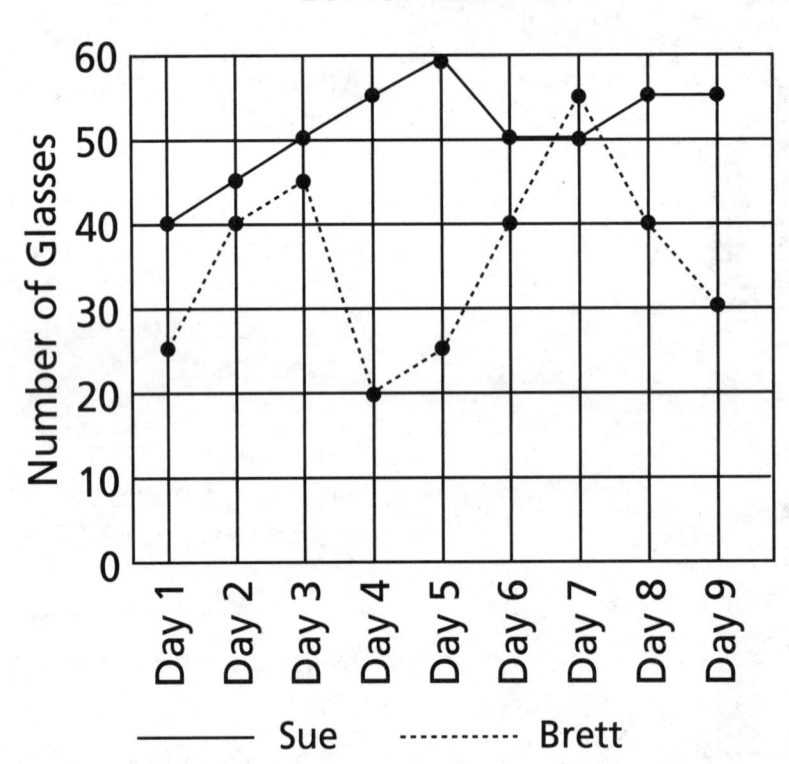

Lemonade Sales

——— Sue   ------- Brett

1. Who sold more lemonade in nine days? by how many more glasses?

   _____

2. In all, how much money did Brett earn?

   _____

3. In all, how much money did Sue earn?

   _____

4. What was the difference in earnings between Sue and Brett?

   _____

5. In the next nine days, if Brett sells four times what he sold this time, how many cups of lemonade will he sell? How much money will he earn?

   _____

6. In the next nine days, if Sue sells half the amount she sold during these nine days, how many cups of lemonade will she sell? How much money will she earn?

   _____

Use with Grade 6, Chapter 5, Lesson 5, pages 120–123.

Name _____

Stem-and-Leaf Plots

P 5-6 PRACTICE

The stem-and-leaf plot shows the number of points scored by an intramural basketball team last season. Use the stem-and-leaf plot for exercises 1–4.

1. How many games did the team play last season?

   _____

2. What was the least number of points scored?

   _____

3. What was the greatest number of points scored?

   _____

4. In how many games did the team score less than 70 points?

   _____

Game Scores

5	2	6	9
6	0	4	6
7	1	5	
8	4	8	
10	7		

Key: 8 | 4   means 84 points

Make a stem-and-leaf plot for each set of data.
Then answer each question.

5. Science test scores:
   83   73   78   60   85
   92   95   85   99   68

6. What is the range of the test scores?

   _____

7. How many test scores were above 75?

   _____

8. Number of pages read for book reports:
   216   190   195   240   232   228
   228   209   244   255   226
   195   205   228   240   250
   200   209   225   257   255

9. What is the range of the data?

   _____

10. How many books had more than 235 pages?

    _____

Use with Grade 6, Chapter 5, Lesson 6, pages 124–125.

29

Name _____

Make an Appropriate Graph

5-7 PRACTICE

Which type of graph would you use to display the data in each table? Explain why. Then make the graph.

1. CDs owned by Patrick

Type of CD	Number of CDs
Country	3
Rock	10
Rap	8
Blues	6
Pop	2

2. The number of laps completed by students jogging around Lincoln Park

Number of Laps	Number of Students
1	4
2	3
3	5
4	2

_____

3. Length of soccer practices for different ages

Age	Number of Minutes
4–6	25
7–9	35
10–12	50
13–15	75

4. Measurement

Time	Temperature
1 P.M.	64°F
2 P.M.	68°F
3 P.M.	70°F
4 P.M.	66°F

_____

Problem Solving
Solve.

5. Write a problem in which you could use a graph to display the data. Share it with others.

_____

30   Use with Grade 6, Chapter 5, Lesson 7, pages 126–127.

Name _____

Range, Mean, Median, and Mode

6-1 PRACTICE

Find the range, mean, median, and mode.
Round to the nearest tenth if necessary.

1. Daily hours of practice before a concert:
   1, 5, 3, 4, 1, 4, 1, 2

   _____

2. One team's hockey scores:
   2, 1, 3, 0, 3, 1, 1, 2, 4, 3

   _____

3. Daily low temperatures (°F) for a week:
   55, 58, 62, 62, 65, 67, 72

   _____

4. Number of seconds for a 200-m run:
   27, 30, 25, 28, 29, 33, 32, 25, 25, 35

   _____

5. Number of grocery purchases of the first ten people in an express check-out lane:
   3, 3, 4, 6, 6, 7, 8, 10, 10, 10

   _____

6. Average miles per gallon:
   20, 20, 20, 20, 20, 20, 20

   _____

7. Daily art show attendance:
   12, 9, 8, 15, 15, 13, 15, 12, 12, 10

   _____

8. Math test scores:
   76, 84, 88, 84, 86, 80, 92, 88, 84, 80, 78, 90

   _____

Problem Solving
Solve.

9. The highest point in Arizona is Mt. Humphreys, at 12,633 ft, and the lowest point, 70 ft, is on the Colorado River. What is the range in elevations?

   _____

10. Paul had scores of 83, 76, 92, 76, and 93 on his math tests. Which of the mean, median, or mode do you think he should use to describe the test scores? Explain.

    _____

    _____

    _____

Use with Grade 6, Chapter 6, Lesson 1, pages 132–134.

Name _____

Choose the Most Appropriate Statistic

6-2 PRACTICE

Find the mean, median, and mode for each set of data. Round to the nearest tenth if necessary. Which of these is the most appropriate average? Explain.

1. 56, 55, 56, 57, 56, 98

2. 12, 13, 23, 24, 11, 25, 21, 14

3. 56, 55, 58, 57, 59, 98

4. Daily craft show attendance: 202, 12, 13, 8, 9, 14, 10, 11, 16, 15, 17

Problem Solving

Solve. State whether n is an outlier in each data set. Write yes or no.

5. 8, n, 7, 9; mean = 13

6. 8, n, 7, 8; mean = 8

7. n, 16, 15, 16, 14; mean = 15

8. 75, n, 76, 78, 77; mean = 104

32 Use with Grade 6, Chapter 6, Lesson 2, pages 136–137.

Name _____

Explore Sampling

P 6-3 PRACTICE

Name each population and sample.

1. Survey each library patron who checks out books to find out if the library should increase its hours of operation.

   _____

   _____

2. Survey shoppers in the frozen-food section at a supermarket to find out which brand of ice cream they prefer.

   _____

   _____

3. Survey sixth graders at the Pembroke School to find out which magazines they read.

   _____

   _____

Decide whether it makes sense to select a sample for the survey. Explain your reasoning.

4. You want to find the favorite magazine of sixth-grade girls.

   _____

   _____

5. You want to find the most popular name for boys in your class.

   _____

   _____

6. You want to determine the most popular car in your city.

   _____

   _____

Describe how you would choose a sample.

7. Survey to find out whether people might go to see a certain movie.

   _____

   _____

8. Survey whether to put a traffic signal at a certain intersection.

   _____

   _____

Use with Grade 6, Chapter 6, Lesson 3, pages 138–139.

Name _____

Sampling

6-4 PRACTICE

Is sample A or sample B more likely to be representative of the population? Explain.

1. About 2,300 people work for Mr. Wilson's company. He wants to know how many people like working there by surveying a sample of the people who work in his company.
   Sample A Put the names of all the people in the company in a hat. Select 200 names from the hat.
   Sample B Put the names of all of the people in the company who have worked there over 5 years in a hat. Select 200 names from the hat.

   _____
   _____
   _____
   _____

2. The principal wants to know which subject the sixth-grade students at Sweet Home Middle School like the best.
   Sample A Put the number of all the sixth-grade classrooms in a bag. Select one number from the bag. Survey all the students in that classroom.
   Sample B Survey the sixth-grade students in the music club.

   _____
   _____
   _____

Use data from the table to answer questions 3–5.

Students Who Exercise Regularly	
Method of Selecting Sample	**Percent**
Carly asked 20 girls in her school.	40%
Jason used a list of all the sixth graders at the school and asked every 10th student on the list.	65%
Rodrigo put the names of all the sixth, seventh, and eighth graders in separate bags and chose 15 students from each bag to ask.	85%

3. Which method is the least likely to be representative of the whole school? _____

4. Which method is the most likely to be representative of the whole school? _____

5. What is the best estimate of the number of students who exercise regularly? _____

Name _____

Problem Solving: Strategy
Do an Experiment

P 6-5 PRACTICE

1. Design an experiment to test the hypothesis that the boys in your class read more mystery books than girls. Tell how you will collect your data and what you will do with the data.

   _____

   _____

2. Do the experiment that you designed in problem 1. Show the data you collected in a table.

   _____

3. What conclusions can you draw from the data you collected in problem 2?

   _____

   _____

4. Present the data you collected in problem 2 in a graph.

   _____

   _____

Mixed Strategy Review
Solve. Use any strategy.

5. Denise is training for a track meet. Today she runs laps around a 1.75-mile hike-and-bike trail. If Denise ran 14 miles, how many laps did she complete?

   _____

   Strategy: _____

6. Edward has 14 times as many small rocks in his collection as large rocks. He has 84 small rocks in his rock collection. How many large rocks does he have?

   _____

   Strategy: _____

7. **Logical Reasoning** Out of 12 students, 9 like to ski, 7 like to snowboard, and 5 like to ski only. How many students like to snowboard only?

   _____

   Strategy: _____

8. There are 9 more dogs in a pet show than cats. There are 25 pets in all. How many cats are in the pet show?

   _____

   Strategy: _____

Use with Grade 6, Chapter 6, Lesson 5, pages 144–145.

Name _____

Divisibility

P 7-1 PRACTICE

Check the divisibility of each number by 2, 3, 4, 5, 6, 8, 9, and 10. List the numbers that it is divisible by.

1. 126 _____
2. 257 _____
3. 430 _____

4. 535 _____
5. 745 _____
6. 896 _____

7. 729 _____
8. 945 _____
9. 4,580 _____

10. 6,331 _____
11. 7,952 _____
12. 8,000 _____

13. 19,450 _____
14. 21,789 _____
15. 43,785 _____

16. 28,751 _____
17. 8,012,120 _____
18. 9,143,001 _____

Which number satisfies the given conditions? Circle A, B, C, or D.

19. divisible by 3 and 5

  A. 10   B. 93   C. 45   D. 54

20. divisible by 2, 3, and 9

  A. 18   B. 9   C. 6   D. 60

21. divisible by 2, 4, 5, 8, and 10

  A. 406   B. 400   C. 205   D. 716

22. divisible by 2, 3, 5, and 10

  A. 708   B. 65   C. 200   D. 600

Problem Solving
Solve.

23. There are 1,224 personal computers on a college campus. There are 8 rooms set aside to keep the computers in. Is it possible for each of these rooms to have the same number of personal computers in them? Explain.

24. Cathy picks 216 pears. She packs the same number of pears into each box. All of the pears are put into boxes. She can choose a box that holds 2, 3, 4, 5, 6, 8, 9, or 10 pears. Which kind of box could she use to pack the pears so that there will be no pears left over?

_____

36   Use with Grade 6, Chapter 7, Lesson 1, pages 160–163.

Name_____

Explore Prime and Composite Numbers

P PRACTICE 7-2

Complete each factor tree. Then write the prime factorization. Use exponents when possible.

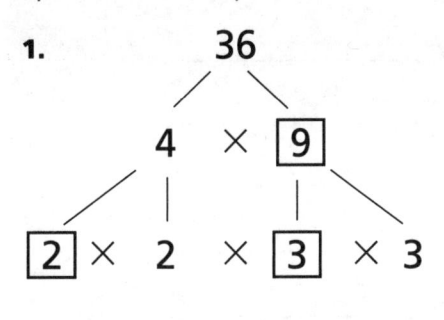

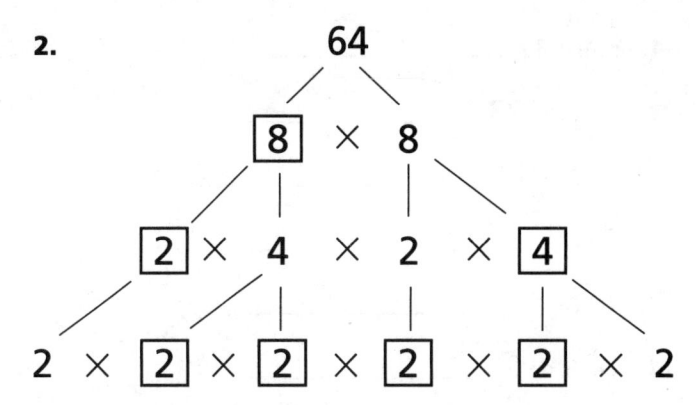

1. 36
    4 × [9]
    [2] × 2 × [3] × 3
    _____

2. 64
    [8] × 8
    [2] × 4 × 2 × [4]
    2 × [2] × [2] × [2] × [2] × 2
    _____

For each number, tell whether it is prime or composite. Write a prime factorization for the composite numbers. Use exponents if you can.

3. 75 _____   4. 61 _____   5. 96 _____
   _____   _____   _____

6. 48 _____   7. 29 _____   8. 95 _____
   _____   _____   _____

9. 68 _____   10. 54 _____   11. 171 _____
   _____   _____   _____

12. 143 _____   13. 117 _____   14. 207 _____
    _____   _____   _____

Problem Solving
Solve.

15. A board is 24 in. long. Find all the whole-number lengths into which it can be evenly divided.

    _____

    _____

16. A ribbon is 36 in. long. Find all the whole-number lengths into which it can be evenly divided.

    _____

    _____

Use with Grade 6, Chapter 7, Lesson 2, pages 164–165.

Greatest Common Factors and Least Common Multiples

7-3 PRACTICE

List all of the common factors.

1. 8 and 12 _____
2. 18 and 27 _____
3. 15 and 23 _____
4. 17 and 34 _____
5. 24 and 12 _____
6. 18 and 24 _____
7. 5 and 25 _____
8. 20 and 25 _____
9. 10 and 15 _____
10. 25 and 75 _____
11. 14 and 21 _____
12. 18 and 19 _____

Find the greatest common factor (GCF).

13. 24 and 28 _____
14. 27 and 36 _____
15. 15 and 305 _____
16. 24 and 45 _____
17. 27 and 57 _____
18. 24 and 48 _____
19. 35 and 56 _____
20. 29 and 87 _____
21. 75 and 200 _____
22. 160 and 900 _____
23. 8, 42, and 60 _____
24. 75, 90, and 120 _____
25. 45, 70, and 120 _____
26. 200, 300, and 450 _____

Find the least common multiple (LCM).

27. 2 and 3 _____
28. 2 and 7 _____
29. 3 and 5 _____
30. 5 and 12 _____
31. 7 and 21 _____
32. 3 and 14 _____
33. 5 and 14 _____
34. 1, 4, and 40 _____
35. 13, 26, and 2 _____
36. 9, 12, and 18 _____

Problem Solving
Solve.

37. The GCF of two numbers is 850. Neither number is divisible by the other. What are the least two numbers these could be?

38. The LCM of two numbers is 51. Both numbers are greater than 1 and both numbers are prime. What are the two numbers?

Name_____

Understanding Fractions

P 7-4 PRACTICE

Name the fraction represented by the shaded part.

1. 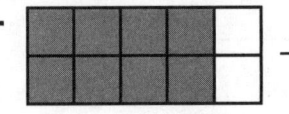 _____   2. _____

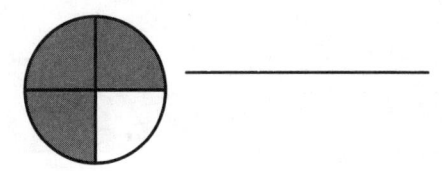

Write two equivalent fractions for each fraction.

3. $\frac{2}{5}$ _____   4. $\frac{6}{18}$ _____   5. $\frac{5}{10}$ _____

6. $\frac{3}{12}$ _____   7. $\frac{21}{35}$ _____   8. $\frac{6}{8}$ _____

9. $\frac{8}{20}$ _____   10. $\frac{3}{9}$ _____   11. $\frac{12}{15}$ _____

12. $\frac{6}{24}$ _____   13. $\frac{12}{20}$ _____   14. $\frac{8}{10}$ _____

Algebra Find the missing number.

15. $\frac{4}{5} = \frac{}{15}$ 16. $\frac{14}{16} = \frac{}{8}$ 17. $\frac{3}{} = \frac{12}{28}$

18. $\frac{2}{6} = \frac{}{3}$ 19. $\frac{}{16} = \frac{3}{4}$ 20. $\frac{3}{4} = \frac{6}{}$

21. $\frac{14}{42} = \frac{}{3}$ 22. $\frac{5}{} = \frac{15}{27}$ 23. $\frac{9}{30} = \frac{3}{}$

Are the fractions equivalent? Write yes or no.

24. $\frac{3}{4} = \frac{6}{8}$ _____   25. $\frac{3}{8} = \frac{7}{16}$ _____   26. $\frac{5}{9} = \frac{15}{27}$ _____

27. $\frac{2}{3} = \frac{4}{5}$ _____   28. $\frac{5}{7} = \frac{15}{21}$ _____   29. $\frac{10}{13} = \frac{7}{14}$ _____

Problem Solving
Solve.

30. Ling ate $\frac{3}{12}$ of a cake. Kai ate $\frac{2}{10}$ of the same cake. Did they eat the same amount of cake? Explain.

_____

_____

31. Jessica ate $\frac{2}{8}$ of a pie that was cut into eighths. John ate the same amount from another pie, exactly the same size, but cut into fourths. How many pieces of his pie did John eat?

_____

Use with Grade 6, Chapter 7, Lesson 4, pages 170–172.

Simplify Fractions

7-5 PRACTICE

Tell whether each fraction is in simplest form. Write yes or no.

1. $\frac{4}{8}$ _____
2. $\frac{25}{100}$ _____
3. $\frac{3}{4}$ _____

4. $\frac{2}{4}$ _____
5. $\frac{4}{5}$ _____
6. $\frac{5}{10}$ _____

7. $\frac{3}{10}$ _____
8. $\frac{15}{20}$ _____
9. $\frac{3}{7}$ _____

10. $\frac{9}{12}$ _____
11. $\frac{14}{20}$ _____
12. $\frac{3}{5}$ _____

13. $\frac{11}{22}$ _____
14. $\frac{3}{51}$ _____
15. $\frac{7}{12}$ _____

16. $\frac{2}{15}$ _____
17. $\frac{9}{21}$ _____
18. $\frac{5}{24}$ _____

19. $\frac{21}{35}$ _____
20. $\frac{15}{50}$ _____
21. $\frac{13}{39}$ _____

22. $\frac{18}{48}$ _____
23. $\frac{19}{57}$ _____
24. $\frac{15}{92}$ _____

Write each fraction in simplest form.

25. $\frac{12}{18}$ _____
26. $\frac{7}{14}$ _____
27. $\frac{3}{25}$ _____
28. $\frac{9}{45}$ _____
29. $\frac{15}{60}$ _____

30. $\frac{7}{21}$ _____
31. $\frac{8}{9}$ _____
32. $\frac{9}{30}$ _____
33. $\frac{3}{15}$ _____
34. $\frac{6}{28}$ _____

35. $\frac{8}{12}$ _____
36. $\frac{12}{15}$ _____
37. $\frac{15}{18}$ _____
38. $\frac{27}{36}$ _____
39. $\frac{20}{24}$ _____

40. $\frac{10}{40}$ _____
41. $\frac{49}{56}$ _____
42. $\frac{13}{52}$ _____
43. $\frac{15}{90}$ _____
44. $\frac{18}{24}$ _____

45. $\frac{36}{82}$ _____
46. $\frac{11}{33}$ _____
47. $\frac{65}{90}$ _____
48. $\frac{28}{63}$ _____
49. $\frac{51}{68}$ _____

50. $\frac{20}{80}$ _____
51. $\frac{31}{93}$ _____
52. $\frac{26}{52}$ _____
53. $\frac{55}{75}$ _____
54. $\frac{88}{88}$ _____

Name _____

Problem Solving: Skill
Extra or Missing Information

7-6 PRACTICE

Use the chart to solve problems 1–2. If there is not enough information to solve, write *not enough information*.

Jeremy is selling tickets to a school concert. He has sold 50 tickets so far.

Day	Number Sold
Thursday	10
Friday	15
Saturday	25

1. What fraction of the tickets sold so far are for Friday?

   _____

2. Jeremy plans to sell 25 more tickets. What fraction of the tickets will be for Thursday?

   _____

Choose the correct answer.

3. Read the problem. Which numbers are needed to solve it?

   In 1999, Lance bicycled 15 miles each day for 8 weeks. For the next 4 weeks, he bicycled 18 miles each day. How many miles did Lance bicycle altogether during the 12 weeks?

   A 12, 15, 18, 4, 8
   B 15, 18, 4, 8
   C 12, 15, 18
   D 1999, 18, 15, 12

4. Read the problem. What additional information do you need to answer the question?

   Tami ran every day last week. She ran 3 miles on Monday, Tuesday, Wednesday, and Friday, and 4 miles on Thursday. How many miles did she run altogether?

   F how fast she ran
   G the number of miles she ran on Saturday
   H the total number of miles she ran on weekdays
   I the total number of miles she ran on the weekend

Mixed Strategy Review

5. Kyra and her friends buy 6 beaded bracelets for $5.95 each. How much did the bracelets cost?

   _____

6. A butterfly sweatshirt costs $18. Butterfly T-shirts, hats, and scarves cost $12 each. How much do 5 hats cost?

   _____

Use with Grade 6, Chapter 7, Lesson 6, pages 176–177.

Name_____

Compare and Order Fractions

P PRACTICE 8-1

Compare. Write >, <, or =.

1. $\frac{1}{2}$ ◯ $\frac{2}{3}$
2. $\frac{2}{9}$ ◯ $\frac{3}{4}$
3. $\frac{6}{8}$ ◯ $\frac{9}{12}$
4. $\frac{7}{10}$ ◯ $\frac{13}{20}$

5. $\frac{1}{5}$ ◯ $\frac{1}{4}$
6. $\frac{1}{6}$ ◯ $\frac{2}{15}$
7. $\frac{5}{6}$ ◯ $\frac{7}{8}$
8. $\frac{4}{5}$ ◯ $\frac{5}{6}$

9. $\frac{5}{8}$ ◯ $\frac{10}{16}$
10. $\frac{1}{2}$ ◯ $\frac{5}{8}$
11. $\frac{2}{3}$ ◯ $\frac{10}{15}$
12. $\frac{2}{3}$ ◯ $\frac{6}{7}$

13. $\frac{3}{8}$ ◯ $\frac{3}{7}$
14. $\frac{6}{7}$ ◯ $\frac{5}{7}$
15. $\frac{1}{6}$ ◯ $\frac{1}{18}$
16. $\frac{3}{10}$ ◯ $\frac{3}{8}$

17. $\frac{4}{5}$ ◯ $\frac{8}{10}$
18. $\frac{2}{7}$ ◯ $\frac{2}{9}$
19. $\frac{3}{10}$ ◯ $\frac{3}{11}$
20. $\frac{2}{5}$ ◯ $\frac{3}{6}$

21. $\frac{5}{6}$ ◯ $\frac{8}{10}$
22. $\frac{2}{7}$ ◯ $\frac{2}{5}$
23. $\frac{5}{11}$ ◯ $\frac{5}{13}$
24. $\frac{3}{5}$ ◯ $\frac{9}{15}$

25. $\frac{3}{12}$ ◯ $\frac{1}{4}$
26. $\frac{4}{8}$ ◯ $\frac{4}{6}$
27. $\frac{8}{10}$ ◯ $\frac{6}{10}$
28. $\frac{5}{6}$ ◯ $\frac{10}{12}$

Order from least to greatest.

29. $\frac{4}{9}, \frac{3}{4}, \frac{5}{6}$ _____
30. $\frac{7}{8}, \frac{9}{10}, \frac{5}{6}$ _____

31. $\frac{1}{3}, \frac{5}{9}, \frac{4}{5}$ _____
32. $\frac{1}{2}, \frac{2}{3}, \frac{4}{9}$ _____

33. $\frac{1}{3}, \frac{5}{9}, \frac{1}{6}$ _____
34. $\frac{3}{5}, \frac{1}{2}, \frac{10}{11}$ _____

35. $\frac{2}{3}, \frac{4}{5}, \frac{1}{2}$ _____
36. $\frac{3}{7}, \frac{1}{3}, \frac{1}{5}$ _____

37. $\frac{9}{12}, \frac{6}{9}, \frac{8}{10}$ _____
38. $\frac{5}{6}, \frac{7}{10}, \frac{3}{8}$ _____

39. $\frac{5}{6}, \frac{1}{6}, \frac{3}{6}$ _____
40. $\frac{3}{5}, \frac{1}{3}, \frac{2}{4}$ _____

41. $\frac{6}{8}, \frac{6}{10}, \frac{6}{9}$ _____
42. $\frac{11}{12}, \frac{4}{7}, \frac{5}{9}$ _____

43. $\frac{5}{6}, \frac{3}{4}, \frac{5}{8}$ _____
44. $\frac{2}{7}, \frac{2}{9}, \frac{2}{3}$ _____

42

Use with Grade 6, Chapter 8, Lesson 1, pages 182–183.

Name _____

Problem Solving: Strategy
Make a Table

8-2 PRACTICE

Use the *make-a-table* strategy to solve.
Use data from the list for problems 1–2.
Write each answer in simplest form.

> **Afternoon Snack** (1 serving)
> Wheat crackers: 120 total calories, 30 fat calories
> Cheese crackers: 140 total calories, 50 fat calories
> Saltine crackers: 60 total calories, 15 fat calories
> Cheddar cheese: 110 total calories, 80 fat calories

1. What fraction of the wheat crackers' calories comes from fat?

2. What fraction of the total calories in one serving of saltine crackers and one serving of cheddar cheese comes from fat?

3. Shelbi made this list of the beads in her collection of beads: triangular: 80 with letters, 200 total; square: 110 with letters, 150 total; circular: 60 with letters, 120 total. What fraction of all her beads have letters?

4. David has 150 baseball cards, 60 basketball cards, and 40 football cards. Of those, 20 are rare baseball cards, 8 are rare basketball cards, and 2 are rare football cards. What fraction of his sports cards are rare sports cards?

Mixed Strategy Review
Solve. Use any strategy.

5. Erin drives 330 miles in 6 hours to visit her cousins. What average speed did she travel?

    Strategy: _____

6. **Time** Owen's digital alarm clock read 9:58 when he fell asleep at night. The alarm went off at 6:50 A.M. but Owen did not get up until 14 minutes later. How long did Owen sleep?

    Strategy: _____

7. **Number Sense** Genevieve has $1.85 in quarters and in dimes only. She has a total of 11 coins. How many of each coin does she have?

    Strategy: _____

8. **Create a problem** for which you could make a table to solve. Share it with others.

Use with Grade 6, Chapter 8, Lesson 2, pages 184–185.

Name_____

Mixed Numbers

P 8-3 PRACTICE

Rename each as an improper fraction.

1. $3\frac{1}{2}$ _____
2. $5\frac{3}{4}$ _____
3. $6\frac{7}{8}$ _____
4. $5\frac{5}{12}$ _____
5. $4\frac{1}{6}$ _____

6. $6\frac{2}{3}$ _____
7. $12\frac{2}{3}$ _____
8. $10\frac{23}{100}$ _____
9. $9\frac{1}{4}$ _____
10. $8\frac{2}{5}$ _____

11. $25\frac{1}{4}$ _____
12. $22\frac{1}{2}$ _____
13. $6\frac{4}{5}$ _____
14. $4\frac{3}{10}$ _____
15. $6\frac{1}{100}$ _____

16. $7\frac{5}{8}$ _____
17. $6\frac{3}{8}$ _____
18. $3\frac{9}{100}$ _____
19. $5\frac{5}{6}$ _____
20. $9\frac{3}{17}$ _____

21. $25\frac{1}{3}$ _____
22. $5\frac{2}{9}$ _____
23. $12\frac{2}{3}$ _____
24. $5\frac{3}{7}$ _____
25. $6\frac{4}{9}$ _____

26. $10\frac{1}{18}$ _____
27. $5\frac{5}{12}$ _____
28. $6\frac{2}{13}$ _____
29. $25\frac{4}{5}$ _____
30. $20\frac{5}{6}$ _____

Rename each as a whole number or mixed number in simplest form.

31. $\frac{11}{3}$ _____
32. $\frac{19}{5}$ _____
33. $\frac{25}{3}$ _____
34. $\frac{42}{6}$ _____
35. $\frac{43}{8}$ _____

36. $\frac{49}{6}$ _____
37. $\frac{36}{4}$ _____
38. $\frac{68}{9}$ _____
39. $\frac{17}{5}$ _____
40. $\frac{13}{7}$ _____

41. $\frac{27}{5}$ _____
42. $\frac{37}{12}$ _____
43. $\frac{21}{4}$ _____
44. $\frac{13}{6}$ _____
45. $\frac{53}{6}$ _____

46. $\frac{43}{9}$ _____
47. $\frac{85}{5}$ _____
48. $\frac{62}{4}$ _____
49. $\frac{31}{3}$ _____
50. $\frac{81}{9}$ _____

Problem Solving
Solve.

51. Tina spent $\frac{10}{3}$ hours practicing piano. Write this quantity as a mixed number.

52. Suppose you have $2\frac{1}{4}$ oranges. Write this quantity as an improper fraction.

_____ _____

Use with Grade 6, Chapter 8, Lesson 3, pages 186–187.

Name_____

Relate Fractions and Decimals

P PRACTICE 8-4

Write the decimal represented by each model. Write this decimal as a fraction in simplest form. Each grid represents 1.

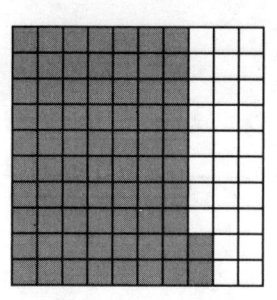

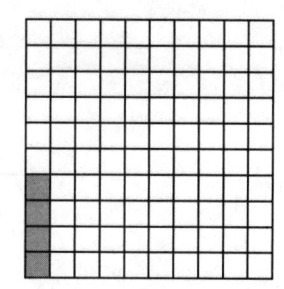

1. _____

2. _____

Write each decimal as a fraction or mixed number in simplest form.

3. 0.6 _____   **4.** 1.25 _____   **5.** 0.74 _____   **6.** 0.29 _____   **7.** 0.635 _____

8. 0.8 _____   **9.** 6.16 _____   **10.** 0.95 _____   **11.** 9.5 _____   **12.** 8.7 _____

13. 2.35 _____   **14.** 0.954 _____   **15.** 0.645 _____   **16** 0.782 _____   **17.** 0.493 _____

Write each fraction or mixed number as a decimal.

18. $\frac{3}{8}$ _____   **19.** $\frac{7}{8}$ _____   **20.** $\frac{9}{16}$ _____   **21.** $2\frac{4}{25}$ _____   **22.** $\frac{9}{100}$ _____

23. $\frac{3}{4}$ _____   **24.** $\frac{7}{25}$ _____   **25.** $\frac{3}{50}$ _____   **26.** $9\frac{1}{8}$ _____   **27.** $5\frac{7}{8}$ _____

28. $11\frac{7}{10}$ _____   **29.** $12\frac{3}{5}$ _____   **30.** $8\frac{5}{20}$ _____   **31.** $4\frac{1}{50}$ _____   **32.** $8\frac{3}{5}$ _____

Problem Solving
Solve.

33. Of the books at the Public Library, $\frac{1}{4}$ are for young readers. What decimal names this fraction?

34. Kathleen has recorded 0.4 of a book on to a cassette tape. What fraction of the book has she recorded?

_____   _____

Use with Grade 6, Chapter 8, Lesson 4, pages 188–190.

45

Compare and Order Fractions and Mixed Numbers

PRACTICE 8-5

Order the numbers from least to greatest.

1. 1.85, $\frac{5}{4}$, $1\frac{3}{8}$ _____

2. 4.1, $4\frac{1}{5}$, $\frac{4}{2}$ _____

3. 0.78, $\frac{7}{8}$, 0.87, $\frac{8}{7}$ _____

4. 0.8, $\frac{3}{4}$, $\frac{5}{8}$ _____

5. $\frac{9}{5}$, $1\frac{3}{4}$, 1.5 _____

6. 1.1, $1\frac{1}{5}$, $\frac{3}{2}$ _____

7. $2\frac{4}{5}$, 2.4, $\frac{10}{4}$ _____

8. 3.2, $\frac{15}{5}$, $3\frac{1}{2}$ _____

9. $\frac{9}{2}$, $4\frac{1}{4}$, 4.3 _____

10. 5.7, $5\frac{3}{5}$, $\frac{15}{2}$ _____

11. $6\frac{2}{5}$, $6\frac{3}{8}$, 6.625 _____

12. $\frac{1}{5}$, $1\frac{1}{12}$, 0.95 _____

13. 3.3, $2\frac{2}{6}$, $\frac{5}{2}$ _____

14. 0.8, $\frac{9}{3}$, $3\frac{3}{4}$ _____

15. $\frac{9}{5}$, $2\frac{3}{5}$, 2.7 _____

16. $\frac{7}{5}$, $1\frac{1}{2}$, $1\frac{3}{4}$ _____

17. $\frac{9}{4}$, $1\frac{3}{5}$, 1.8 _____

18. $\frac{3}{5}$, 3.1, $3\frac{2}{10}$ _____

19. $\frac{5}{9}$, $\frac{1}{4}$, 0.5 _____

20. $\frac{8}{4}$, 1.75, $1\frac{2}{7}$ _____

Problem Solving
Solve.

21. Harrison bought 1.45 pounds of peanuts to add to a trail mix. Was the weight of the peanuts greater or less than $1\frac{1}{2}$ pounds?

_____

22. Marta bought 1.6 pounds of dried fruit. A trail mix recipe calls for $1\frac{3}{4}$ pounds of dried fruit. Does Marta have enough for the recipe?

_____

Add and Subtract Fractions with Like Denominators

9-1 PRACTICE

Add or subtract. Write your answer in simplest form.

1. $\frac{6}{7} - \frac{3}{7}$

2. $\frac{3}{4} + \frac{1}{4}$

3. $\frac{7}{10} - \frac{1}{10}$

4. $\frac{4}{5} - \frac{3}{5}$

5. $\frac{7}{12} - \frac{5}{12}$

6. $\frac{10}{11} + \frac{2}{11}$

7. $\frac{3}{5} + \frac{4}{5}$

8. $\frac{8}{9} - \frac{5}{9}$

9. $\frac{5}{8} + \frac{1}{8}$

10. $\frac{5}{6} - \frac{1}{6}$

11. $\frac{9}{15} - \frac{6}{15}$

12. $\frac{11}{12} + \frac{5}{12}$

13. $\frac{7}{8} + \frac{5}{8}$

14. $\frac{5}{6} - \frac{3}{6}$

15. $\frac{9}{10} - \frac{3}{10}$

16. $\frac{4}{9} - \frac{1}{9} =$

17. $\frac{7}{15} + \frac{2}{15} =$

18. $\frac{5}{6} + \frac{3}{6} =$

19. $\frac{7}{10} + \frac{3}{10} =$

20. $\frac{2}{9} + \frac{8}{9} =$

21. $\frac{6}{12} + \frac{2}{12} =$

22. $\frac{4}{5} - \frac{1}{5} =$

23. $\frac{3}{8} + \frac{3}{8} =$

24. $\frac{7}{9} - \frac{2}{9} =$

25. $\frac{5}{6} - \frac{2}{6} =$

Problem Solving

Solve. Write your answer in simplest form.

26. Jayne needs $\frac{7}{8}$ of a yard of ribbon to decorate a banner. She has $\frac{5}{8}$ of a yard of ribbon. How much more ribbon does Jayne need?

27. Manuel walked $\frac{2}{3}$ of a mile to the park. He walked the same distance back home. How far did Manuel walk altogether?

Use with Grade 6, Chapter 9, Lesson 1, pages 208–210.

Name _____

Explore Adding and Subtracting Fractions with Unlike Denominators

9-2 PRACTICE

Use fraction models to find each sum or difference.
Write your answer in simplest form.

1. $\frac{1}{3} - \frac{1}{6} =$ _____
2. $\frac{1}{2} + \frac{1}{8} =$ _____
3. $\frac{5}{6} + \frac{1}{3} =$ _____
4. $\frac{9}{10} - \frac{4}{5} =$ _____
5. $\frac{1}{4} + \frac{3}{8} =$ _____
6. $\frac{1}{2} - \frac{1}{4} =$ _____
7. $\frac{5}{8} - \frac{1}{4} =$ _____
8. $\frac{7}{8} - \frac{1}{4} =$ _____
9. $\frac{1}{6} + \frac{1}{2} =$ _____
10. $\frac{3}{10} + \frac{1}{5} =$ _____
11. $\frac{1}{2} - \frac{5}{12} =$ _____
12. $\frac{7}{9} - \frac{1}{3} =$ _____
13. $\frac{1}{4} + \frac{5}{6} =$ _____
14. $\frac{5}{6} - \frac{1}{3} =$ _____
15. $\frac{4}{9} + \frac{1}{3} =$ _____
16. $\frac{3}{4} - \frac{1}{6} =$ _____
17. $\frac{5}{6} - \frac{1}{4} =$ _____
18. $\frac{3}{8} + \frac{3}{4} =$ _____

Problem Solving
Solve.

19. Lin and Ron ate $\frac{1}{3}$ of a pizza for lunch. They ate $\frac{1}{2}$ of the pizza for dinner. How much of the pizza did they eat in all? Use models to solve.

    _____

20. Ellen sold $\frac{3}{7}$ of her boxes of cookies before lunch and $\frac{1}{2}$ of the boxes after lunch. What fraction of her boxes of cookies did she sell that day?

    _____

48   Use with Grade 6, Chapter 9, Lesson 2, pages 212–213.

Name _____

Add Fractions with Unlike Denominators

P 9-3 PRACTICE

Add. Write your answer in simplest form.

1. $\frac{2}{3} + \frac{3}{5}$
2. $\frac{2}{3} + \frac{5}{9}$
3. $\frac{3}{4} + \frac{5}{8}$
4. $\frac{2}{7} + \frac{5}{14}$
5. $\frac{1}{2} + \frac{5}{6}$
6. $\frac{11}{12} + \frac{3}{4}$

7. $\frac{5}{12} + \frac{1}{4}$
8. $\frac{7}{15} + \frac{1}{6}$
9. $\frac{8}{9} + \frac{2}{3}$
10. $\frac{5}{6} + \frac{3}{8}$
11. $\frac{7}{15} + \frac{1}{3}$
12. $\frac{3}{4} + \frac{3}{10}$

13. $\frac{2}{9} + \frac{5}{6}$
14. $\frac{4}{5} + \frac{3}{4}$
15. $\frac{11}{12} + \frac{7}{8}$
16. $\frac{7}{10} + \frac{1}{6}$
17. $\frac{7}{8} + \frac{2}{3}$
18. $\frac{9}{10} + \frac{9}{15}$

19. $\frac{2}{5} + \frac{7}{10} =$ _____
20. $\frac{5}{6} + \frac{4}{9} =$ _____
21. $\frac{2}{3} + \frac{1}{4} =$ _____

22. $\frac{7}{10} + \frac{1}{5} =$ _____
23. $\frac{3}{4} + \frac{1}{3} =$ _____
24. $\frac{5}{6} + \frac{2}{9} =$ _____

25. $\frac{1}{2} + \frac{3}{10} =$ _____
26. $\frac{1}{2} + \frac{7}{8} =$ _____
27. $\frac{5}{8} + \frac{1}{2} =$ _____

28. $\frac{3}{8} + \frac{5}{6} =$ _____
29. $\frac{3}{4} + \frac{2}{5} =$ _____
30. $\frac{3}{5} + \frac{1}{4} =$ _____

31. $\frac{2}{5} + \frac{3}{10} =$ _____
32. $\frac{3}{4} + \frac{2}{3} =$ _____
33. $\frac{3}{10} + \frac{3}{4} =$ _____

Problem Solving

Solve.

34. Cathy spent $\frac{2}{5}$ of an hour on her French assignment and $\frac{4}{5}$ of an hour on her English report. How much time did she spend doing her homework? Write your answer in simplest form.

_____

35. On Saturday, Jason spent $\frac{1}{5}$ of his time skateboarding and $\frac{1}{10}$ of his time reading. What fraction of his time did Jason spend skateboarding and reading?

_____

Use with Grade 6, Chapter 9, Lesson 3, pages 214–216.

Subtract Fractions with Unlike Denominators

P 9-4 PRACTICE

Subtract. Write your answer in simplest form.

1. $\frac{2}{3} - \frac{3}{5}$
2. $\frac{2}{3} - \frac{5}{9}$
3. $\frac{3}{4} - \frac{5}{8}$
4. $\frac{5}{7} - \frac{5}{14}$
5. $\frac{1}{2} - \frac{1}{6}$
6. $\frac{11}{12} - \frac{3}{4}$

7. $\frac{5}{12} - \frac{1}{4}$
8. $\frac{7}{15} - \frac{1}{6}$
9. $\frac{8}{9} - \frac{2}{3}$
10. $\frac{5}{6} - \frac{3}{8}$
11. $\frac{7}{15} - \frac{1}{3}$
12. $\frac{3}{4} - \frac{4}{10}$

13. $\frac{8}{9} - \frac{5}{6}$
14. $\frac{4}{5} - \frac{3}{4}$
15. $\frac{11}{12} - \frac{7}{8}$
16. $\frac{7}{10} - \frac{1}{6}$
17. $\frac{7}{4} - \frac{5}{8}$
18. $\frac{9}{10} - \frac{9}{15}$

19. $\frac{4}{5} - \frac{7}{10} =$ _____
20. $\frac{5}{6} - \frac{4}{9} =$ _____
21. $\frac{2}{3} - \frac{1}{4} =$ _____
22. $\frac{7}{10} - \frac{1}{5} =$ _____
23. $\frac{3}{4} - \frac{1}{3} =$ _____
24. $\frac{5}{6} - \frac{2}{9} =$ _____
25. $\frac{1}{2} - \frac{3}{10} =$ _____
26. $\frac{1}{2} - \frac{3}{8} =$ _____
27. $\frac{5}{8} - \frac{1}{2} =$ _____
28. $\frac{3}{8} - \frac{1}{6} =$ _____
29. $\frac{3}{4} - \frac{2}{5} =$ _____
30. $\frac{3}{5} - \frac{1}{4} =$ _____
31. $\frac{2}{5} - \frac{1}{6} =$ _____
32. $\frac{3}{4} - \frac{2}{3} =$ _____
33. $\frac{9}{10} - \frac{3}{4} =$ _____

Problem Solving
Solve.

34. Clifton spent $\frac{2}{3}$ hour practicing guitar. He spent $\frac{1}{6}$ hour changing the strings on his guitar. How much longer did he spend practicing?

35. In the new den, $\frac{1}{6}$ of the walls will be made of glass blocks, and $\frac{1}{8}$ will be covered with tile. The rest of the den will be covered with wood paneling. What fraction of the room will be covered with glass blocks and tile?

Use with Grade 6, Chapter 9, Lesson 4, pages 218–219.

Name _____

Problem Solving: Skill
Multistep Problems

9-5 PRACTICE

Use data from the list for problems 1–3.
Organizers of an art fair have created a list showing the use of space for various exhibits.

1. Landscapes will use $\frac{1}{4}$ of the total space and still lifes will use $\frac{1}{6}$ of the total space. What fraction of the space is left?

   _____

2. What fraction of the total space is left to be set up after the abstract paintings and the prints are set up?

   _____

3. What fraction of the total space will have to be set up after the prints and portraits are set up?

   _____

Art Exhibits	Fraction of Total Space
Landscapes	$\frac{1}{4}$
Still Lifes	$\frac{1}{6}$
Portraits	$\frac{1}{3}$
Abstracts	$\frac{1}{12}$
Watercolors	$\frac{1}{8}$
Prints	$\frac{1}{24}$

Choose the correct answer.

4. Brad bought tickets to the fair on Friday. He used $\frac{1}{8}$ of the tickets on Friday and $\frac{1}{2}$ of the tickets on Saturday. What fraction of the tickets did he have left for Sunday?

   A $\frac{1}{8}$
   B $\frac{1}{4}$
   C $\frac{3}{8}$
   D $\frac{5}{8}$

5. Sondra used $\frac{1}{3}$ of her tokens playing games and $\frac{2}{5}$ of her tokens on rides at the amusement park. What fraction of the tokens does she have left?

   F $\frac{3}{8}$
   G $\frac{4}{15}$
   H $\frac{5}{8}$
   I $\frac{11}{15}$

Mixed Strategy Review

6. Malcolm has $2.30 in quarters and dimes. He has two more dimes than quarters. How many quarters does Malcolm have?

   _____

7. In a display of model cars, each row has two more model cars than the one before it. The first row has 3 model cars. How many model cars are in 5 rows?

   _____

Name_____

Add Mixed Numbers

P 10-1
PRACTICE

Add. Write your answer in simplest form.

1. $5\frac{2}{3}$
   $+ 3\frac{3}{4}$

2. $12\frac{7}{8}$
   $+ 4\frac{1}{4}$

3. $13\frac{1}{2}$
   $+ 4\frac{3}{5}$

4. $21\frac{1}{3}$
   $+ 5\frac{7}{24}$

5. $8\frac{1}{2}$
   $+ 6\frac{4}{5}$

6. $5\frac{3}{8}$
   $+ 6\frac{11}{12}$

7. $5\frac{1}{5}$
   $+ 2\frac{1}{3}$

8. $9\frac{2}{5}$
   $+ 8\frac{3}{8}$

9. $4\frac{1}{6}$
   $+11\frac{1}{2}$

10. $7\frac{7}{8}$
    $+ 1\frac{11}{12}$

11. $4\frac{3}{10} + 5\frac{2}{5} = $ _____

12. $3\frac{7}{8} + 2\frac{1}{2} = $ _____

13. $5\frac{2}{3} + 3\frac{1}{4} = $ _____

14. $6\frac{3}{4} + 2\frac{1}{2} = $ _____

15. $1\frac{1}{12} + 3\frac{1}{6} = $ _____

16. $9\frac{2}{5} + 10\frac{3}{10} = $ _____

17. $7\frac{1}{3} + 5\frac{11}{12} = $ _____

18. $11\frac{7}{10} + 4 = $ _____

19. $2\frac{2}{3} + 4\frac{3}{4} = $ _____

20. $7\frac{3}{4} + 2\frac{7}{8} = $ _____

21. $4\frac{1}{2} + 3\frac{5}{6} = $ _____

22. $7\frac{2}{3} + 1\frac{5}{6} = $ _____

23. $2\frac{1}{4} + 4\frac{3}{5} = $ _____

24. $5\frac{3}{8} + 7\frac{1}{4} = $ _____

25. $14\frac{5}{16} + 8\frac{3}{8} = $ _____

26. $15\frac{3}{4} + 12\frac{5}{8} = $ _____

27. $9\frac{7}{8} + 4\frac{5}{6} = $ _____

28. $12\frac{7}{8} + 6\frac{1}{3} = $ _____

Problem Solving
Solve.

29. A cave is $5\frac{1}{2}$ miles west of a waterfall. A group of hikers is $2\frac{1}{4}$ miles east of the waterfall. How far is the group of hikers from the cave?

_____

30. A mark on the side of a pier shows that the water is $4\frac{7}{8}$ ft deep. When the tide is high, the depth increases by $2\frac{3}{4}$ ft. What is the depth of the water when the tide is high?

_____

52

Use with Grade 6, Chapter 10, Lesson 1, pages 226–227.

Name _____

Explore Subtracting Mixed Numbers

P 10-2 PRACTICE

Complete. Cross out fraction pieces to find $2\frac{1}{4} - 1\frac{1}{2}$.

1. $2\frac{1}{4} - 1\frac{1}{2} = 2\frac{1}{4} - 1\frac{\square}{\square}$

   $= 1\frac{\square}{4} - 1\frac{2}{4}$

   $= \frac{\square}{\square}$

| 1 | 1 | $\frac{1}{4}$ |

↓

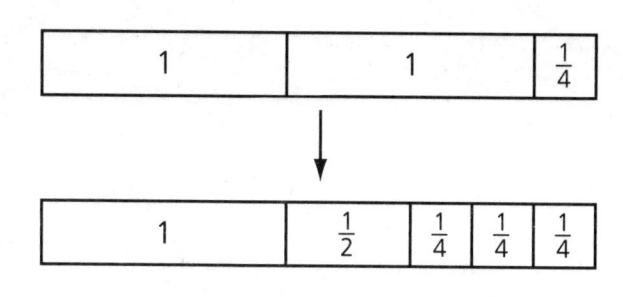

| 1 | $\frac{1}{2}$ | $\frac{1}{4}$ | $\frac{1}{4}$ | $\frac{1}{4}$ |

Use models to find the difference.

2. $2\frac{1}{8} - 1\frac{5}{8} =$ _____

3. $3\frac{2}{3} - 2\frac{1}{6} =$ _____

4. $3\frac{7}{12} - 1\frac{11}{12} =$ _____

5. $7 - 3\frac{7}{12} =$ _____

6. $4\frac{1}{3} - 2\frac{2}{3} =$ _____

7. $6\frac{1}{4} - 4\frac{3}{4} =$ _____

8. $3 - 1\frac{1}{2} =$ _____

9. $4\frac{1}{2} - 2\frac{3}{8} =$ _____

10. $7\frac{1}{2} - 5\frac{2}{3} =$ _____

11. $12 - 4\frac{3}{8} =$ _____

12. $7\frac{1}{10} - \frac{9}{10} =$ _____

13. $13 - 4\frac{4}{5} =$ _____

14. $7\frac{5}{6} - 3\frac{1}{4} =$ _____

15. $12\frac{1}{10} - 4\frac{3}{10} =$ _____

16. $11\frac{3}{8} - 6\frac{3}{4} =$ _____

17. $14\frac{3}{5} - 6\frac{7}{10} =$ _____

18. $15\frac{7}{8} - 9\frac{1}{3} =$ _____

19. $17\frac{4}{5} - 8\frac{3}{4} =$ _____

Problem Solving
Solve.

20. A large table is $30\frac{3}{4}$ in. high. A small table is $16\frac{5}{16}$ in. high. How much higher is the larger table?

21. Branda is $59\frac{1}{2}$ inches tall. Her sister is $48\frac{3}{4}$ inches tall. How much taller is Branda than her sister?

Use with Grade 6, Chapter 10, Lesson 2, pages 228–229.

Name_____

Subtract Mixed Numbers

10-3 PRACTICE

Subtract. Write your answer in simplest form.

1. $10\frac{11}{16} - 3\frac{7}{8}$

2. $8\frac{1}{3} - 2\frac{3}{8}$

3. $9 - 3\frac{2}{5}$

4. $5\frac{3}{16} - 2\frac{3}{8}$

5. $8\frac{1}{6} - 3\frac{2}{5}$

6. $7\frac{1}{2} - 3$

7. $2\frac{3}{4} - 1\frac{1}{8}$

8. $4\frac{1}{8} - 2\frac{1}{16}$

9. $9\frac{2}{3} - 3\frac{5}{6}$

10. $2\frac{1}{10} - 1\frac{2}{5}$

11. $15\frac{7}{12} - 8\frac{1}{2} =$ _____

12. $6\frac{7}{16} - 2\frac{7}{8} =$ _____

13. $27\frac{1}{4} - 13\frac{11}{12} =$ _____

14. $5\frac{2}{5} - 1\frac{1}{4} =$ _____

15. $10\frac{2}{3} - 7\frac{3}{4} =$ _____

16. $7\frac{1}{4} - 2\frac{5}{6} =$ _____

17. $8\frac{1}{2} - 1\frac{2}{3} =$ _____

18. $10\frac{1}{2} - 2\frac{4}{5} =$ _____

19. $12\frac{2}{3} - 6\frac{3}{4} =$ _____

20. $5\frac{1}{2} - 3\frac{3}{4} =$ _____

21. $15\frac{1}{8} - 7\frac{3}{4} =$ _____

22. $11\frac{1}{4} - 6\frac{5}{8} =$ _____

Algebra Find each missing number.

23. $4\frac{1}{5} + \square = 6\frac{3}{10}$

24. $8\frac{7}{12} + \square = 15\frac{1}{3}$

25. $6\frac{5}{8} + \square = 10\frac{2}{3}$

26. $\square + 2\frac{3}{7} = 6\frac{1}{35}$

Problem Solving
Solve.

27. Anna has $3\frac{1}{4}$ yd of fabric. She plans to use $2\frac{1}{2}$ yd for curtains. Does she have enough left to make 2 pillows that each use $\frac{5}{8}$ yd of fabric? Explain.

28. Paula has 2 yd of elastic. One project needs a $\frac{3}{4}$-yd piece. Does she have enough for another project that needs $1\frac{1}{3}$ yd? Explain.

54   Use with Grade 6, Chapter 10, Lesson 3, pages 230–232.

Name _____

Problem Solving: Strategy
Find a Pattern

P 10-4 PRACTICE

1. Jesse is increasing the amount of weight he lifts each week over time. During the first four weeks he lifted $3\frac{1}{2}$, $4\frac{1}{4}$, 5, and $5\frac{3}{4}$ pounds. Based on his pattern, how much weight will he lift during the fifth week? _____

2. Ivory rides her bike each day for nine days. The first five days she rode $9\frac{5}{8}$, $9\frac{1}{4}$, $8\frac{7}{8}$, $8\frac{1}{2}$, and $8\frac{1}{8}$ miles. Based on her pattern, how long will she ride on the ninth day? _____

3. Miguel jogs each day for one week. During the first four days he jogs for $\frac{1}{3}$, $\frac{1}{2}$, $\frac{2}{3}$, and $\frac{5}{6}$ hour. Based on his pattern, how long will he jog on the seventh day? _____

4. The number of miles Avania skates each day for five days is $8\frac{1}{3}$, $7\frac{2}{3}$, 7, $6\frac{1}{3}$, and $5\frac{2}{3}$ miles. Based on her pattern, how many miles will she skate on the eighth day? _____

Mixed Strategy Review
Solve. Use any strategy.

5. Design an experiment to test the hypothesis that girls in your class play more board games than boys. Tell how you will collect your data and what you will do with the data.

   _____

   Strategy: _____

6. Hal earns $735.25 testing video games each month. After three months, he received a bonus of $250 for recommending a new game. How much did Hal earn for all three months?

   _____

   Strategy: _____

7. Anita has three times as many books as Nellie and 10 fewer books than Clara. Clara has 100 books. How many books does Nellie have?

   _____

   Strategy: _____

8. **Write a problem** that can be solved by finding a pattern. Share it with others.

   _____

   _____

Use with Grade 6, Chapter 10, Lesson 4, pages 234–235.

Name _____

Estimate Sums and Differences

P 10-5 PRACTICE

Estimate the sum or difference by rounding.

1. $\frac{1}{6} + \frac{5}{8}$ _____
2. $\frac{7}{8} - \frac{1}{16}$ _____
3. $\frac{9}{10} + \frac{7}{8}$ _____

4. $8\frac{1}{12} + 5\frac{9}{10}$ _____
5. $6\frac{1}{10} + 5\frac{5}{6}$ _____
6. $10\frac{4}{5} - \frac{1}{6}$ _____

7. $20\frac{11}{12} - 5\frac{5}{16}$ _____
8. $15\frac{15}{16} + 5\frac{11}{12}$ _____
9. $2\frac{1}{6} + 7\frac{1}{9}$ _____

10. $4\frac{9}{10} - 3\frac{5}{8}$ _____
11. $4\frac{7}{8} + 8\frac{1}{5}$ _____
12. $14\frac{7}{9} - 9\frac{1}{8}$ _____

13. $14\frac{3}{4} + 9\frac{7}{8}$ _____
14. $7\frac{11}{15} - 6\frac{7}{16}$ _____
15. $3\frac{11}{15} - 2\frac{9}{10}$ _____

16. $8\frac{7}{8} - \frac{11}{12}$ _____
17. $8\frac{5}{12} + 25\frac{1}{15}$ _____
18. $10\frac{3}{5} - 5\frac{1}{8}$ _____

19. $12\frac{7}{8} + 22\frac{7}{9}$ _____
20. $24\frac{7}{8} - 4\frac{5}{9}$ _____
21. $7\frac{4}{5} + 2\frac{1}{3}$ _____

22. $1\frac{7}{8} + 3\frac{2}{5}$ _____
23. $4\frac{1}{8} - 1\frac{8}{9}$ _____
24. $8\frac{9}{16} - 3\frac{7}{8}$ _____

25. $34\frac{1}{2} - 20\frac{1}{4}$ _____
26. $72\frac{2}{5} - 44\frac{1}{8}$ _____
27. $28\frac{1}{20} - 10\frac{7}{10}$ _____

28. $59\frac{2}{3} + 80\frac{1}{4}$ _____
29. $16\frac{4}{5} - 9\frac{1}{3}$ _____
30. $28\frac{9}{10} - 12\frac{1}{5}$ _____

31. $45\frac{3}{4} + 54\frac{1}{8}$ _____
32. $35\frac{13}{15} - 15\frac{3}{10}$ _____
33. $82\frac{17}{20} - 12\frac{1}{7}$ _____

Algebra Estimate the value of each expression using $n = 2\frac{4}{5}$.

34. $n + 3\frac{1}{3}$ _____
35. $25 - n$ _____
36. $16\frac{6}{7} - n$ _____

Problem Solving
Solve.

37. John's car can hold $16\frac{1}{10}$ gallons of gasoline. About how many gallons are left if he starts with a full tank and uses $11\frac{9}{10}$ gallons?

38. Julia bought stock at $\$28\frac{1}{8}$ per share. The value of each stock increased by $\$6\frac{5}{8}$. About how much is each share of stock now worth?

_____   _____

56   Use with Grade 6, Chapter 10, Lesson 5, pages 236–237.

Name _____

Fractions of Whole Numbers

P 11-1 PRACTICE

Multiply. Use mental math.

1. $\frac{1}{5}$ of 20 _____
2. $\frac{3}{5}$ of 20 _____
3. $\frac{1}{2}$ of 14 _____
4. $\frac{1}{4} \times 28 =$ _____
5. $\frac{3}{4} \times 28 =$ _____
6. $\frac{1}{3} \times 15 =$ _____
7. $\frac{2}{3} \times 15 =$ _____
8. $\frac{1}{8} \times 56 =$ _____
9. $\frac{1}{4} \times 40 =$ _____
10. $\frac{2}{5} \times 10 =$ _____
11. $\frac{3}{7} \times 21 =$ _____
12. $\frac{1}{6} \times 42 =$ _____
13. $\frac{1}{10} \times 30 =$ _____
14. $\frac{1}{3} \times 27 =$ _____
15. $\frac{1}{5} \times 25 =$ _____
16. $\frac{2}{3} \times 63 =$ _____
17. $\frac{2}{9} \times 72 =$ _____
18. $\frac{2}{3} \times 24 =$ _____
19. $\frac{5}{6} \times 72 =$ _____
20. $\frac{4}{9} \times 45 =$ _____
21. $\frac{2}{3} \times 33 =$ _____
22. $\frac{3}{5} \times 55 =$ _____
23. $\frac{2}{3} \times 90 =$ _____
24. $\frac{3}{5} \times 50 =$ _____
25. $\frac{1}{8} \times 240 =$ _____
26. $\frac{2}{7} \times 280 =$ _____
27. $\frac{1}{11} \times 22 =$ _____
28. $\frac{1}{2} \times 160 =$ _____
29. $\frac{2}{3} \times 900 =$ _____
30. $\frac{3}{5} \times 500 =$ _____
31. $\frac{1}{9} \times 720 =$ _____
32. $\frac{1}{6} \times 300 =$ _____
33. $\frac{7}{9} \times 810 =$ _____

Algebra Find the missing number. Explain which operation you used, and why you used it.

34. $\frac{1}{2} \times n = 30$ _____
35. $\frac{1}{3} \times a = 12$ _____
36. $\frac{1}{6} \times b = 200$ _____
37. $n \times 25 = 5$ _____
38. $a \times 200 = 50$ _____
39. $s \times 54 = 9$ _____
40. $n \times 63 = 14$ _____
41. $b \times 250 = 75$ _____
42. $\frac{5}{8} \times n = 25$ _____

Problem Solving
Solve.

43. Cathy surveyed 54 students and found that $\frac{5}{9}$ of them prefer vanilla ice cream. How many students prefer vanilla ice cream?

44. Matilda surveyed 95 students and found that $\frac{4}{5}$ of them prefer pepperoni as a pizza topping. How many students prefer pepperoni as a pizza topping?

Use with Grade 6, Chapter 11, Lesson 1, pages 252–253.

Name _____

Explore Multiplying Fractions

P 11-2
PRACTICE

Tell how many rows and columns the model of each product should have.

1. $\frac{1}{2} \times \frac{2}{3}$ _____

2. $\frac{1}{3} \times \frac{2}{5}$ _____

3. $\frac{1}{3} \times \frac{3}{4}$ _____

4. $\frac{1}{5} \times \frac{1}{4}$ _____

5. $\frac{3}{4} \times \frac{1}{2}$ _____

6. $\frac{2}{3} \times \frac{1}{4}$ _____

7. $\frac{1}{5} \times \frac{1}{2}$ _____

8. $\frac{1}{3} \times \frac{4}{5}$ _____

Use grid paper to find each product. Write your answer in simplest form.

9. $\frac{1}{2} \times \frac{2}{3} =$ _____

10. $\frac{1}{3} \times \frac{2}{5} =$ _____

11. $\frac{1}{3} \times \frac{3}{4} =$ _____

12. $\frac{1}{5} \times \frac{1}{4} =$ _____

13. $\frac{3}{4} \times \frac{1}{2} =$ _____

14. $\frac{2}{3} \times \frac{1}{4} =$ _____

15. $\frac{1}{5} \times \frac{1}{2} =$ _____

16. $\frac{1}{3} \times \frac{4}{5} =$ _____

17. $\frac{1}{2} \times \frac{3}{5} =$ _____

18. $\frac{1}{4} \times \frac{1}{3} =$ _____

19. $\frac{4}{5} \times \frac{1}{2} =$ _____

20. $\frac{2}{3} \times \frac{1}{5} =$ _____

21. $\left(\frac{1}{3}\right)^2 =$ _____

22. $\left(\frac{3}{4}\right)^2 =$ _____

23. $\left(\frac{2}{5}\right)^2 =$ _____

24. $\frac{2}{5} \times \frac{3}{5} =$ _____

25. $\frac{3}{4} \times \frac{2}{3} =$ _____

26. $\frac{1}{3} \times \frac{2}{3} =$ _____

Name_____

Multiply Fractions

P 11-3 PRACTICE

Multiply. Write your answer in simplest form.

1. $\frac{5}{7} \times \frac{3}{20} =$ _____
2. $\frac{4}{5} \times \frac{3}{8} =$ _____
3. $\frac{9}{10} \times \frac{2}{3} =$ _____

4. $\frac{2}{3} \times \frac{4}{5} =$ _____
5. $\frac{3}{4} \times \frac{1}{5} =$ _____
6. $\frac{3}{8} \times \frac{4}{9} =$ _____

7. $\frac{5}{12} \times \frac{3}{4} =$ _____
8. $\frac{3}{5} \times \frac{4}{7} =$ _____
9. $\frac{2}{3} \times 15 =$ _____

10. $\frac{3}{5} \times \frac{10}{21} =$ _____
11. $\frac{6}{15} \times \frac{5}{18} =$ _____
12. $9 \times \frac{2}{3} =$ _____

13. $\frac{14}{25} \times \frac{5}{7} =$ _____
14. $\frac{7}{12} \times \frac{6}{11} =$ _____
15. $5 \times \frac{15}{25} =$ _____

16. $\frac{3}{5} \times \frac{10}{12} =$ _____
17. $\frac{1}{4} \times \frac{4}{5} =$ _____
18. $\frac{1}{3} \times \frac{3}{5} =$ _____

19. $\frac{1}{5} \times \frac{5}{12} =$ _____
20. $4 \times \frac{1}{10} =$ _____
21. $\frac{1}{4} \times \frac{8}{12} =$ _____

22. $\frac{1}{2} \times \frac{8}{12} =$ _____
23. $3 \times \frac{2}{3} =$ _____
24. $\frac{3}{4} \times \frac{4}{5} =$ _____

25. $\frac{2}{3} \times \frac{3}{10} =$ _____
26. $4 \times \frac{4}{8} =$ _____
27. $\frac{3}{4} \times \frac{5}{6} =$ _____

28. $\frac{7}{8} \times \frac{3}{5} =$ _____
29. $\frac{3}{5} \times \frac{5}{6} =$ _____
30. $\frac{2}{5} \times \frac{3}{8} =$ _____

Algebra Evaluate each expression using the value given.

31. $\frac{1}{2}n$, for $n = \frac{3}{4}$ _____
32. $\frac{3}{5}x$, for $x = \frac{5}{16}$ _____

33. $\frac{1}{5}p$, for $p = \frac{4}{5}$ _____
34. $\frac{5}{12}c$, for $c = \frac{11}{15}$ _____

35. $\frac{2}{3}k$, for $k = \frac{2}{3}$ _____
36. $\frac{3}{4}h$, for $h = \frac{3}{4}$ _____

37. $\frac{5}{9}z$, for $z = \frac{3}{5}$ _____
38. $\frac{7}{10}q$, for $q = \frac{5}{14}$ _____

Problem Solving
Solve.

39. A town plans to use one half of a square for a picnic area. One third of the picnic area will be a playground. What portion of the square will be a playground?

40. Suppose one fourth of the playground is for baseball. What portion of the square is for baseball?

_____

Use with Grade 6, Chapter 11, Lesson 3, pages 256–258.

Name_____

Multiply Mixed Numbers

P PRACTICE 11-4

Multiply. Write your answer in simplest form.

1. $2\frac{5}{6} \times 1\frac{3}{4} =$ _____
2. $3\frac{3}{8} \times 7\frac{1}{4} =$ _____
3. $5\frac{3}{8} \times 2\frac{7}{8} =$ _____

4. $2\frac{3}{8} \times 4\frac{4}{5} =$ _____
5. $6\frac{7}{12} \times 5\frac{9}{10} =$ _____
6. $7\frac{1}{3} \times 10\frac{11}{12} =$ _____

7. $12\frac{1}{4} \times 3\frac{3}{4} =$ _____
8. $8\frac{1}{6} \times 2\frac{1}{4} =$ _____
9. $15\frac{2}{3} \times 5\frac{5}{7} =$ _____

10. $\frac{1}{4} \times 5\frac{2}{5} =$ _____
11. $2\frac{3}{8} \times \frac{4}{5} =$ _____
12. $1\frac{1}{2} \times 5\frac{1}{3} =$ _____

13. $3\frac{3}{8} \times 6 =$ _____
14. $\frac{3}{4} \times 1\frac{3}{5} =$ _____
15. $9\frac{3}{5} \times \frac{1}{3} =$ _____

16. $1\frac{1}{4} \times 2\frac{2}{3} =$ _____
17. $1\frac{3}{5} \times \frac{1}{4} =$ _____
18. $6\frac{1}{4} \times 1\frac{2}{5} =$ _____

19. $\frac{7}{8} \times 3\frac{1}{5} =$ _____
20. $5\frac{1}{3} \times 2\frac{1}{4} =$ _____
21. $\frac{3}{5} \times 4\frac{1}{2} =$ _____

22. $\frac{5}{8} \times 7\frac{3}{5} =$ _____
23. $6\frac{1}{4} \times 7\frac{1}{2} =$ _____
24. $3\frac{1}{4} \times 2\frac{2}{3} =$ _____

25. $9\frac{3}{4} \times 2\frac{2}{5} =$ _____
26. $4\frac{1}{5} \times 5\frac{2}{3} =$ _____
27. $1\frac{1}{5} \times 1\frac{7}{8} =$ _____

28. $3\frac{1}{4} \times 2\frac{5}{6} =$ _____
29. $1\frac{1}{2} \times 2\frac{1}{2} =$ _____
30. $2\frac{3}{4} \times 3\frac{1}{4} =$ _____

31. $5\frac{1}{3} \times \frac{5}{8} =$ _____
32. $2\frac{4}{5} \times \frac{3}{7} =$ _____
33. $3\frac{1}{3} \times 3\frac{3}{10} =$ _____

34. $5\frac{1}{2} \times \frac{2}{5} =$ _____
35. $4\frac{3}{4} \times 1\frac{1}{8} =$ _____
36. $7\frac{5}{9} \times 4\frac{2}{3} =$ _____

37. $2\frac{1}{8} \times 2\frac{2}{3} =$ _____
38. $2\frac{1}{4} \times 1\frac{2}{3} =$ _____
39. $3\frac{2}{5} \times 2\frac{1}{2} =$ _____

Problem Solving

Solve.

40. Ken used a piece of lumber to build a bookshelf. If he made three shelves that are each $2\frac{1}{2}$ ft long, how long was the piece of lumber?

41. Deanna's cake recipe calls for $2\frac{1}{4}$ cups of flour. She needs to double the recipe for a bake sale. How much flour should Deanna use?

_____

_____

Use with Grade 6, Chapter 11, Lesson 4, pages 260–261.

Name _____

Problem Solving: Skill
Choose the Operation

11-5 PRACTICE

Solve. Tell what operation(s) you used.

1. A department store receives a delivery of 540 glasses. A clerk finds that $\frac{1}{10}$ of the glasses are chipped. How many glasses are chipped?

   _____

2. A chef needs to have 2 gallons of sauce available to use for dinner orders. She has $\frac{3}{4}$ gallon of the sauce already made. How much more sauce does she need?

   _____

3. A carpenter has 36 boards that are each 8 inches long. Half of the boards are stained with a sealant. What is the total length of the boards in feet that are stained with a sealant?

   _____

4. Vicky collects key chains. Of the 48 key chains in her collection, $\frac{1}{3}$ of them are cartoon characters. How many key chains in her collection are not cartoon characters?

   _____

Choose the correct answer.

Last year, Calvin's time in a 10-mile race was $1\frac{3}{4}$ hours. This year, his time in the same race was $\frac{1}{4}$ hour less than his time last year.

5. Which of the following statements is true?

   A Calvin ran slower this year than last year.
   B Calvin had a better race time last year than this year.
   C Calvin ran the race in $\frac{1}{4}$ of the time this year.
   D Calvin's time improved this year over last year's time.

6. To find Calvin's time for this year, you should

   F add $1\frac{3}{4}$ and $\frac{1}{4}$.
   G subtract $\frac{1}{4}$ from $1\frac{3}{4}$.
   H multiply $1\frac{3}{4}$ by $\frac{1}{4}$.
   J divide $1\frac{3}{4}$ by $\frac{1}{4}$.

Mixed Strategy Review

7. Greg has a $1 bill, a quarter, and a dime. How many different total amounts of money can he make?

   _____

8. A seed company packaged 55,000 packets of seeds. Of the packets, $\frac{3}{5}$ have been sold. How many packets have *not* been sold?

   _____

Use with Grade 6, Chapter 11, Lesson 5, pages 262–263.

Name_____

Estimate Products and Quotients

P 12-1 PRACTICE

Estimate the product. Show your method.

1. $\frac{3}{5} \times 9\frac{4}{5}$ = _____
2. $\frac{1}{3} \times 17\frac{2}{3}$ = _____
3. $\frac{3}{8} \times 24\frac{2}{7}$ = _____
4. $6\frac{1}{3} \times 12\frac{1}{7}$ = _____
5. $4\frac{2}{3} \times 3\frac{3}{4}$ = _____
6. $3\frac{1}{5} \times 6\frac{1}{2}$ = _____
7. $2\frac{1}{2} \times 3\frac{4}{9}$ = _____
8. $7\frac{5}{8} \times 9\frac{3}{5}$ = _____
9. $7\frac{5}{8} \times 13\frac{4}{5}$ = _____
10. $\frac{4}{5} \times 9\frac{2}{7}$ = _____

Estimate the quotient. Show your method.

11. $24\frac{2}{3} \div 2\frac{2}{9}$ = _____
12. $15\frac{1}{9} \div 4\frac{1}{8}$ = _____
13. $26\frac{2}{5} \div 8\frac{6}{7}$ = _____
14. $13\frac{1}{2} \div 1\frac{7}{8}$ = _____
15. $14\frac{1}{3} \div 2\frac{5}{6}$ = _____
16. $10\frac{3}{8} \div 3\frac{1}{5}$ = _____
17. $8\frac{2}{3} \div 1\frac{5}{6}$ = _____
18. $24\frac{3}{4} \div 3\frac{2}{7}$ = _____
19. $17\frac{1}{6} \div 5\frac{7}{8}$ = _____
20. $19\frac{4}{5} \div 4\frac{7}{10}$ = _____

Multiply. Estimate to check for reasonableness. Show your estimate.

21. $11\frac{1}{2} \times \frac{2}{3}$ = _____
22. $4\frac{1}{4} \times \frac{3}{8}$ = _____
23. $1\frac{3}{4} \times \frac{5}{6}$ = _____
24. $\frac{3}{5} \times 15\frac{1}{2}$ = _____

Problem Solving
Solve.

25. Carol works $7\frac{1}{2}$ hours a day. Last week she worked $5\frac{3}{4}$ days. About how many hours did she work?

26. Nathan needs ribbon pieces that are $5\frac{1}{4}$ inches long for his mobile. Estimate how many pieces he can cut from a ribbon that is $26\frac{1}{2}$ inches long.

62

Use with Grade 6, Chapter 12, Lesson 1, pages 268–270.

Explore Dividing by Fractions

P 12-2 PRACTICE

Write the division sentence that each diagram represents.
Find each answer.

1.

1		1
$\frac{1}{3}$ $\frac{1}{3}$ $\frac{1}{3}$		$\frac{1}{3}$ $\frac{1}{3}$ $\frac{1}{3}$

_____

2.

1	1	1
$\frac{1}{4}$ $\frac{1}{4}$ $\frac{1}{4}$ $\frac{1}{4}$	$\frac{1}{4}$ $\frac{1}{4}$ $\frac{1}{4}$ $\frac{1}{4}$	$\frac{1}{4}$ $\frac{1}{4}$ $\frac{1}{4}$ $\frac{1}{4}$

_____

3.

1
$\frac{1}{5}$ $\frac{1}{5}$ $\frac{1}{5}$ $\frac{1}{5}$ $\frac{1}{5}$

_____

4.

1	1
$\frac{1}{8}$ $\frac{1}{8}$ $\frac{1}{8}$ $\frac{1}{8}$ $\frac{1}{8}$ $\frac{1}{8}$ $\frac{1}{8}$ $\frac{1}{8}$	$\frac{1}{8}$ $\frac{1}{8}$ $\frac{1}{8}$ $\frac{1}{8}$ $\frac{1}{8}$ $\frac{1}{8}$ $\frac{1}{8}$ $\frac{1}{8}$

_____

Use fraction models to divide. Record your work.

5. $4 \div \frac{1}{3} =$ _____   **6.** $2 \div \frac{1}{6}$ _____   **7.** $3 \div \frac{1}{2} =$ _____

8. $1 \div \frac{1}{9} =$ _____   **9.** $5 \div \frac{1}{2} =$ _____   **10.** $3 \div \frac{1}{5} =$ _____

Divide.

11. $\frac{2}{3} \div \frac{1}{3} =$ _____   **12.** $4 \div \frac{1}{6} =$ _____   **13.** $5 \div \frac{1}{4} =$ _____

Use with Grade 6, Chapter 12, Lesson 2, pages 272–273.

Name _____

Divide Fractions and Mixed Numbers

P 12-3 PRACTICE

Complete.

1. $\frac{2}{5} \div \frac{1}{8} = \frac{2}{5} \times$ ___ = ___ = ___
2. $\frac{3}{4} \div \frac{3}{7} = \frac{3}{4} \times$ ___ = ___ = ___
3. $\frac{3}{5} \div 5 = \frac{3}{5} \times$ ___ = ___
4. $\frac{2}{3} \div \frac{3}{19} = \frac{2}{3} \times$ ___ = ___ = ___

Divide. Write your answer in simplest form. Check by multiplying.

5. $\frac{3}{10} \div \frac{4}{5} =$ _____
6. $\frac{3}{8} \div 3 =$ _____
7. $3 \div 1\frac{4}{5} =$ _____

8. $2\frac{1}{5} \div 1\frac{5}{6} =$ _____
9. $1\frac{1}{2} \div \frac{3}{16} =$ _____
10. $\frac{1}{4} \div \frac{1}{8} =$ _____

11. $1\frac{7}{8} \div \frac{5}{8} =$ _____
12. $1\frac{3}{4} \div \frac{1}{16} =$ _____
13. $3 \div \frac{3}{8} =$ _____

14. $\frac{4}{5} \div \frac{4}{7} =$ _____
15. $\frac{7}{8} \div \frac{7}{9} =$ _____
16. $6\frac{1}{2} \div 2\frac{1}{6} =$ _____

17. $5\frac{1}{3} \div 2\frac{2}{3} =$ _____
18. $6\frac{1}{4} \div 2\frac{1}{2} =$ _____
19. $10 \div 3\frac{1}{3} =$ _____

20. $2\frac{7}{10} \div 4\frac{4}{5} =$ _____
21. $4\frac{2}{3} \div 1\frac{3}{4} =$ _____
22. $5\frac{1}{8} \div 2\frac{1}{2} =$ _____

Compare. Write >, <, or =.

23. $\frac{3}{4} \div \frac{2}{3}$ ◯ $\frac{2}{3} \div \frac{3}{4}$
24. $\frac{1}{2} \div \frac{2}{5}$ ◯ $\frac{2}{5} \div \frac{1}{2}$
25. $3 \div \frac{1}{2}$ ◯ $\frac{1}{2} \div 3$
26. $\frac{1}{5} \div 6$ ◯ $6 \div \frac{1}{5}$

Evaluate. Write the answer in simplest form.

27. $(\frac{1}{2} \times \frac{4}{5}) \div \frac{3}{4} =$ _____
28. $(\frac{2}{7} \times \frac{4}{9}) \div \frac{1}{3} =$ _____

Problem Solving
Solve.

29. Joe had a piece of string $8\frac{3}{4}$ ft long. He cut it into small pieces. Each piece is $1\frac{3}{4}$ ft long. How many pieces did he cut?

30. Brenda had $5\frac{1}{4}$ qt of paint. She used the paint for 3 art projects. She used the same amount of paint for each art project. How much paint did she use for each art project?

Name _____

Problem Solving: Strategy
Work Backward

P 12-4 PRACTICE

Work backward to solve.

1. Mr. McCoy made new business cards. He put $\frac{1}{2}$ of them in his planner and 8 of them in his wallet. He gave 5 of them to his business partner. If he keeps 12 of the cards at home, how many business cards were there to begin with?

   _____

2. The Montgomerys are sending invitations to a family reunion. Wanda sent $\frac{1}{2}$ of the invitations. Her brother sent 7 invitations and her cousin sent the remaining 11 invitations. How many invitations were sent altogether?

   _____

3. Of the 48 magazines in a health club, $\frac{5}{8}$ are fitness magazines. Only $\frac{3}{10}$ of the fitness magazines focus on using weights. How many magazines in the club focus on using weights?

   _____

4. Janna spent $\frac{1}{2}$ of her allowance on a new book. With the remaining money she spent $1 on a new pen and put $3 in savings. How much was Janna's allowance?

   _____

Mixed Strategy Review
Solve. Use any strategy.

5. The first five numbers in a pattern are 512, 256, 128, 64, and 32. What is the tenth number in the pattern? The eleventh number?

   _____

   Strategy: _____

6. Lisa earns $5.50 for each lawn she mows. During the summer, she mows 2 lawns 3 days a week. How much will she earn if she mows lawns for 16 weeks?

   _____

   Strategy: _____

7. Conrad wants to watch a show at 7:30 P.M. He needs 1 hour to do his homework, 30 minutes to eat dinner, and 45 minutes to complete his chores. What is the latest time that he can start his homework in order to watch the show?

   _____

   Strategy: _____

8. **Write a problem** that you could work backward to solve. Share it with others.

   _____

   _____

   _____

Use with Grade 6, Chapter 12, Lesson 4, pages 278–279.

Name _____

Time

13-1 PRACTICE

Complete.

1. 4 weeks = _____ days
2. 180 s = _____ min
3. 100 years = _____ decades

4. 3 days = _____ h
5. 4 min = _____ s
6. 5 h = _____ min

7. 10 min = _____ s
8. 36 weeks = _____ d
9. 120 s = _____ min

10. 500 years = _____ centuries
11. 2 h = _____ s
12. 2 centuries = _____ decades

13. 30 min = _____ s
14. 3 days = _____ min
15. 4 weeks = _____ h

16. 250 s = _____ min _____ s
17. 160 min = _____ h _____ min

18. 78 h = _____ days _____ h
19. 375 min = _____ h _____ min

Find the elapsed time.

20. From 7:45 P.M. to 11:10 P.M.

21. From 10:30 P.M. to 3:20 A.M.

22. From 7:05 A.M. to 8:03 A.M.

23. From 9:48 A.M. to 12:40 P.M.

24. From 10:10 P.M. to 6:45 A.M.

25. From 8:25 A.M. to 9:32 P.M.

Problem Solving

Solve.

26. Lara started the hike to the summit at 7:55 A.M. She reached the summit at 1:05 P.M. How long did it take Lara to reach the summit?

27. There are 366 days in a leap year. How many hours are there in a leap year?

Name _____

Customary Length

P 13-2 PRACTICE

Estimate and then measure the length of each object. Find the measurement to the nearest $\frac{1}{4}$ inch or $\frac{1}{8}$ inch as shown.

1. to the nearest $\frac{1}{4}$ in.

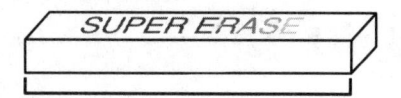

SUPER ERASE

Estimate: _____

Measurement: _____

2. to the nearest $\frac{1}{8}$ in.

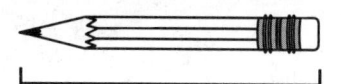

Estimate: _____

Measurement: _____

Choose an appropriate unit for measuring each length. Write *in.*, *ft*, *yd*, or *mi*.

3. length of a classroom _____

4. length of a pencil _____

5. distance between two cities _____

6. length of a football field _____

7. thickness of a book _____

8. width of Atlantic Ocean _____

Complete.

9. 4 ft = _____ in.

10. $1\frac{1}{2}$ yd = _____ in.

11. 15 ft = _____ yd

12. 3 yd = _____ in.

13. 5 mi = _____ ft

14. 40 in. = _____ ft

15. 180 in. = _____ yd

16. $2\frac{1}{3}$ yd = _____ ft

17. 4 mi = _____ yd

18. $\frac{1}{2}$ mi = _____ ft

19. $5\frac{1}{2}$ ft = _____ in.

20. 5 ft 11 in. = _____ in.

21. 16 ft = _____ yd _____ ft

22. 75 in. = _____ ft _____ in.

Problem Solving

Solve.

23. Marjorie is $4\frac{3}{4}$ feet tall. How many inches tall is she?

_____

24. The jogging track at the park is $\frac{1}{8}$ mile long. How many feet long is the track?

_____

Use with Grade 6, Chapter 13, Lesson 2, pages 298–301.

Name _____

Customary Capacity and Weight

P PRACTICE 13-3

Complete.

1. 4,000 lb = _____ T
2. 3 qt = _____ pt
3. 40 oz = _____ lb

4. 4 c = _____ fl oz
5. $6\frac{1}{2}$ lb = _____ oz
6. $5\frac{1}{2}$ qt = _____ pt

7. $\frac{7}{8}$ lb = _____ oz
8. 36 qt = _____ gal
9. 4 qt = _____ fl oz

10. $\frac{3}{4}$ T = _____ lb
11. 128 oz = _____ lb
12. 64 fl oz = _____ pt

13. 3 c = _____ fl oz
14. $4\frac{1}{2}$ T = _____ lb
15. 72 fl oz = _____ c

16. 3,200 lb = _____ T _____ lb
17. 70 fl oz = _____ qt _____ fl oz

18. 75 oz = _____ lb _____ oz
19. 34 pt = _____ gal _____ qt

Compare. Write >, < or =.

20. 20 qt ◯ 4 gal 2 pt
21. 8 pt 1 c ◯ 2 gal
22. 50 fl oz ◯ $1\frac{1}{2}$ qt

23. 3,000 lb ◯ $1\frac{1}{2}$ T
24. $\frac{1}{2}$ T ◯ 1,000 oz
25. 100 oz ◯ 6 lb 3 oz

Problem Solving
Solve.

26. Robert needs 3 pints of milk to make a casserole. He has 5 cups of milk. How many more cups of milk does Robert need?

27. Maria's dog weighs 240 ounces. How many pounds does her dog weigh?

_____

68

Use with Grade 6, Chapter 13, Lesson 3, pages 302–304.

Name_____

Problem Solving: Skill
Choose an Appropriate Tool and Unit

P 13-4 PRACTICE

Choose an appropriate unit of measurement.

1. The length of a bridge.

   _____

2. The weight of a lion.

   _____

For each situation, explain how you would measure.

3. Alan wants to know if he has enough orange juice to make a smoothie.

   _____
   _____

4. A veterinarian wants to know how much a baby panda bear has gained during one week.

   _____
   _____

Choose the correct answer.

5. Which tool would you use to measure the height of a tree?

   A tape measure

   B scale

   C measuring cup

   D thermometer

6. Which tool should a coach use to measure how fast a swimmer swims 10 laps?

   F scale

   G tape measure

   H stopwatch

   J thermometer

Mixed Strategy Review
Solve. Use any strategy.

7. Matthew has 128 pints of oil to recycle. He wants to transport the oil in 1-gallon containers. How many containers will he need?

   _____

8. Lake Superior is 531 feet deeper than Lake Ontario. Lake Ontario is 592 feet deeper than Lake Erie. Lake Erie is 210 feet deep. How deep is Lake Superior?

   _____

Use with Grade 6, Chapter 13, Lesson 4, pages 306–307.

69

Name _____

Metric Length

P 14-1 PRACTICE

Estimate the length of each line segment. Then find its length to the nearest centimeter or millimeter as shown.

1. to the nearest centimeter.

_____

Estimate: _____

Measurement: _____

2. to the nearest millimeter

_____

Estimate: _____

Measurement: _____

Complete.

3. 40 mm = _____ cm

4. 10 km = _____ m

5. 5.36 mm = _____ cm

6. 21,250 m = _____ km

7. 8,560 mm = _____ m

8. 340 m = _____ km

9. 0.05 km = _____ m

10. 7.6 km = _____ m

11. 62 mm = _____ cm

12. 0.45 m = _____ cm

13. 50 m = _____ km

14. 18 m = _____ mm

15. 63 mm = _____ cm

16. 92.1 mm = _____ cm

17. 7,099 m = _____ km

18. 206 cm = _____ m

19. 49,500 m = _____ km

20. 8.9 m = _____ cm

Compare. Write >, <, or =.

21. 590 mm ◯ 5.9 m

22. 310 m ◯ 3,100 cm

23. 4,800 cm ◯ 48 m

24. 0.7 km ◯ 700 m

25. 5,200 mm ◯ 52 cm

26. 0.6 m ◯ 600 cm

Problem Solving
Solve.

27. Ken is 1.8 m tall. His sister is 1.48 m tall. How many centimeters taller is Ken than his sister? Explain.

_____

_____

28. The length of Amanda's cat excluding its tail is 23 centimeters. The length of the cat's tail is 190 millimeters. How many centimeters long is Amanda's cat? Explain.

_____

_____

Use with Grade 6, Chapter 14, Lesson 1, pages 312–314.

Name _____

Metric Mass and Capacity

P 14-2 PRACTICE

Complete.

1. 7,200 mL = _____ L
2. 490 mL = _____ L
3. 90 g = _____ kg
4. 0.1 L = _____ mL
5. 300 g = _____ kg
6. 7.88 mL = _____ L
7. 0.9 kg = _____ g
8. 5 g = _____ kg
9. 0.0042 kg = _____ g
10. 3 L = _____ mL
11. 25 kg = _____ g
12. 8 mL = _____ L
13. 9,260 mL = _____ L
14. 670 g = _____ kg
15. 0.53 L = _____ mL

Compare. Write >, <, or =.

16. 2.4 g ◯ 240 mg
17. 8 kg ◯ 80,000 g
18. 2.32 L ◯ 2,320 mL
19. 410 mL ◯ 4.1 L
20. 3,500 mg ◯ 35 g
21. 970 mL ◯ 9.7 L

Circle the best estimate of the capacity or mass of each object.

22. a soccer ball
    a. 400 g
    b. 4 g
    c. 0.4 g

23. a bicycle
    a. 5 kg
    b. 50 kg
    c. 500 kg

24. a fish tank for a home
    a. 0.4 L
    b. 40 L
    c. 4,000 L

25. a water bottle
    a. 100 L
    b. 1 L
    c. 0.1 L

26. a feather
    a. 10 g
    b. 0.1 g
    c. 100 g

Problem Solving
Solve.

27. A box of pasta has a mass of 454 grams. How many boxes should Leo buy if he wants to cook at least 1 kilogram of pasta? Explain.

28. Tracy has a 5-liter punch bowl. She buys two containers of juice that hold 1.75 liters and 2.75 liters. Can she empty the two containers into the bowl? Explain.

Use with Grade 6, Chapter 14, Lesson 2, pages 316–318.

Name _____

Problem Solving: Strategy
Draw a Diagram

14-3 PRACTICE

Draw a diagram to solve.

1. Jacob is not as tall as Cara. Brittany is taller than Cara but not as tall as Dennis. Which of the four friends is the tallest?

   _____

2. Quinn left his house and rode 3 miles east and then 2 miles south to the library. From there he rode 1 mile west and 4 miles north to Henri's house. Henri and Quinn rode 1 mile south and 2 miles west to the park. How far north was Quinn from his house?

   _____

3. A builder is building 40-meter-wide houses along one side of a 0.3-kilometer-long street. Each house must be at least 20 meters from the house next to it. Also, the two end houses must be at least 20 meters from the edge of the property. What is the greatest number of houses the builder can build?

   _____

4. An archaeologist wants to enclose a rectangular area that is 20 meters long and 15 meters wide for further study. She will need a stake planted every 5 meters to build the enclosure. How many stakes will she need?

   _____

Mixed Strategy Review
Solve. Use any strategy.

5. For a cheese sandwich, Mark can choose whole wheat, white, multigrain, or rye bread. His choices of cheese are Jack, cheddar, and Swiss. How many different sandwiches with 1 kind of bread and 1 kind of cheese are possible?

   _____

   Strategy: _____

6. The length of a rectangular window is 3 times as long as the width. What is the distance around of the window if the width is 8 inches?

   _____

   Strategy: _____

7. Maddie read a book in three days. The first day she read 58 pages. The second day she read twice as many pages as the first day. The third day there were 34 pages fewer than the second day left to read. How long was the book?

   _____

   Strategy: _____

8. **Write a problem** for which you could draw a diagram to solve. Share it with others.

   _____

   _____

   _____

72   Use with Grade 6, Chapter 14, Lesson 3, pages 320–321.

Name _____

Explore Conversions Between Systems

P 14-4 PRACTICE

Measure the following items in different units. Record your measurements.

1. length of paper clip _____ in. _____ cm
2. length of pencil _____ in. _____ cm
3. length of eraser _____ in. _____ cm
4. length of index card _____ in. _____ cm
5. length of classroom _____ yd _____ m
6. width of chalkboard _____ yd _____ m
7. height of a table _____ yd _____ m
8. length of a window _____ yd _____ m

Complete. Round to the nearest tenth.

9. 100 mi ≈ _____ km
10. 6 in. ≈ _____ cm
11. 100 cm ≈ _____ in.
12. 5 kg ≈ _____ lb
13. 1.5 L ≈ _____ c
14. 120 lb ≈ _____ kg
15. 6 c ≈ _____ L
16. 100 m ≈ _____ in.
17. 200 mi ≈ _____ km
18. 2.5 m ≈ _____ ft
19. 30 oz ≈ _____ g
20. 60 in. ≈ _____ cm
21. $\frac{1}{4}$ L ≈ _____ c
22. 10 kg ≈ _____ lb
23. 20 L ≈ _____ qt
24. 60 in. ≈ _____ m
25. 100 km ≈ _____ mi
26. 450 oz ≈ _____ kg
27. $\frac{3}{4}$ m ≈ _____ ft
28. 120 lb ≈ _____ kg
29. 21 m ≈ _____ yd

Problem Solving
Solve.

30. Sarah's stack of quarters measures 3 inches. Maria's stack of quarters measures 9 centimeters. Who has more money? Explain.

_____

_____

31. Kim's drinking bottle holds 1.5 L of water. Ling's bottle holds 2 qt of water. Whose bottle holds more water? Explain.

_____

_____

32. Ben's cat is 5 lb. Tomas's cat is 5 kg. Whose cat has the lesser mass? Explain.

_____

_____

33. Sam ran 5 km yesterday. Julie ran 5 mi. Who ran farther? Explain.

_____

_____

Use with Grade 6, Chapter 14, Lesson 4, pages 322–323.

Temperature

P 14-5 PRACTICE

Write each temperature in degrees Celsius. Round to the nearest degree.

1. 80°F _____
2. 65°F _____
3. 42°F _____
4. 95°F _____
5. 32°F _____
6. 57°F _____
7. 54°F _____
8. 98°F _____
9. 89°F _____
10. 46°F _____
11. 36°F _____
12. 63°F _____
13. 105°F _____
14. 109°F _____
15. 67°F _____

Write each temperature in degrees Fahrenheit. Round to the nearest degree.

16. 40°C _____
17. 12°C _____
18. 0°C _____
19. 95°C _____
20. 27°C _____
21. 20°C _____
22. 30°C _____
23. 37°C _____
24. 25°C _____
25. 10°C _____
26. 7°C _____
27. 36°C _____
28. 98°C _____
29. 6°C _____
30. 50°C _____

Answer each question.

31. What is the temperature in degrees Fahrenheit when it is 0°C? _____

32. What is the temperature in degrees Fahrenheit when it is 100°C? _____

33. What is the temperature in degrees Celsius when it is 41°F? _____

34. What is the temperature in degrees Celsius when it is 77°F? _____

Problem Solving
Solve.

35. The normal body temperature is 98.6°F. Does a temperature of 38.5°C indicate a fever? Explain.

36. The average temperature in Quito, Ecuador, is 57°F. What is the average temperature in degrees Celsius, rounded to the nearest degree?

74   Use with Grade 6, Chapter 14, Lesson 5, pages 324–325.

Name _____

Order of Operations

P 15-1 PRACTICE

Evaluate using the order of operations.

1. $6.5 - 2.4 \div 6 \times 2$ _____
2. $33 - (2^2 + 3^2)$ _____
3. $4 \times 3 + 20 \div 5$ _____
4. $26 + 5^2 - 4 \times 2$ _____
5. $400 \div (44 - 24)$ _____
6. $45 \div 9 + 6 \times 3$ _____
7. $0.7 \times (4 + 6) \times 0.3$ _____
8. $13 + 5 \times 12 - 4$ _____

Evaluate each expression.

9. $24 \div d$ for $d = 2^2$ _____
10. $\frac{p}{2} + 8$ for $p = 6$ _____
11. $3r - 2$ for $r = 65$ _____
12. $8 - 3y$ for $y = 0.4$ _____
13. $6n - (m + 8)$ for $m = 20$ and $n = 15$ _____
14. $5x - y$ for $x = 12$ and $y = 14$ _____
15. $5d - (h + 9)$ for $d = 3$ and $h = 5$ _____
16. $x^2 + 2y$ for $x = 3$ and $y = 2$ _____

Rewrite with parentheses to make each sentence true.

17. $12 + 6 \div 2 - 1 = 8$ _____
18. $14 \div 2 + 5 - 1 = 1$ _____
19. $1 + 2 \times 15 - 4 = 33$ _____
20. $11 - 7 \div 2 = 2$ _____
21. $14 - 3 - 2 \times 3 = 11$ _____
22. $5 \times 6 \div 2 + 1 = 10$ _____

Problem Solving
Solve.

23. Kevin has $0.60. If Dennis has $0.25 more than four times as much money as Kevin does, how much money does Dennis have?

    _____

24. Allison walks to school. Gene rides his bike to school. Allison can walk to school in 15 minutes. If it takes Gene 10 minutes less than twice Allison's time to get to school, how long does it take Gene to ride his bike to school?

    _____

Use with Grade 6, Chapter 15, Lesson 1, pages 340–341.

Name _____

Functions

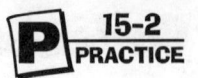

P 15-2 PRACTICE

Complete the table. Write an equation to show the relationship.

1.
Input	x	0	1	2	3	4	5
Output	y	1	3	5	7		

_____

2.
Input	x	0	1	2	3	4	5
Output	y	3	4	5	6		

_____

3.
Input	x	0	1	2	3	4	5
Output	y	1	4	7	10		

_____

4.
Input	x	0	1	2	3	4	5
Output	y	5	7	9	11		

_____

Write an equation for the function described in words. Tell what each variable in the equation represents.

5. The width of a certain rectangle is $\frac{1}{4}$ its length.

_____

_____

6. The length of a certain rectangle is $2\frac{1}{2}$ times its width.

_____

_____

7. The length in inches of a pencil is equal to 2.54 times its length in centimeters.

_____

_____

Sequences and Functions • Algebra

P 15-3 PRACTICE

Explain the rule. Find the next number in the sequence.

1. 5, 11, 17, 23, 29, …

2. 1, 3, 9, 27, 81, …

3. 2, 3, 5, 8, 12, …

4. 1, $\frac{1}{2}$, $\frac{1}{4}$, $\frac{1}{8}$, $\frac{1}{16}$, …

5. 2, 14, 26, 38, 50, …

6. 8, 16, 32, 64, 128, …

7. 800, 400, 200, 100, 50, …

8. 45, 42, 39, 36, 33, …

9. 5, 20, 80, 320, 1,280, …

10. 1, 1.75, 2.5, 3.25, 4, …

Explain the rule. Find the stated number.

11. 8th term in 3, 9, 15, 21, 27, …

12. 8th term in 2, 11, 20, 29, 38, …

13. 7th term in $\frac{1}{4}$, $\frac{1}{2}$, 1, 2, 4, …

14. 7th term in 4, 5, 7, 10, 14, …

Algebra Explain the rule. Find the missing number.

15. 12, _____, 20, 24, 28

16. _____, 14, 28, 56, 112

17. 10, 35, 60, _____, 110

18. 243, 81, _____, 9, 3

Use with Grade 6, Chapter 15, Lesson 3, pages 344–346.

77

Graph a Function

15-4 PRACTICE

Complete the table. Then graph the function.

1. y = x

x	1	2	3	4
y	1			

2. y = x + 2

x	0	1	2	3
y	2			

3. y = 3x

x	0	1	2	3
y				

4. y = 2x − 1

x	1	2	3	4
y	1	3	5	7

5. y = x + 3

x	1	3	5	7
y				

6. y = 2x + 3

x	0	1	2	3
y				

The graph at the right shows the amount of fuel a car uses for different distances traveled.

Use the graph to find the number of gallons of gasoline used for each distance.

7. 90 mi _____
8. 210 mi _____
9. 120 mi _____
10. 270 mi _____
11. 330 mi _____
12. 300 mi _____
13. 60 mi _____
14. 150 mi _____
15. 360 mi _____
16. 180 mi _____

17. How many miles per gallon of gasoline does this car get?

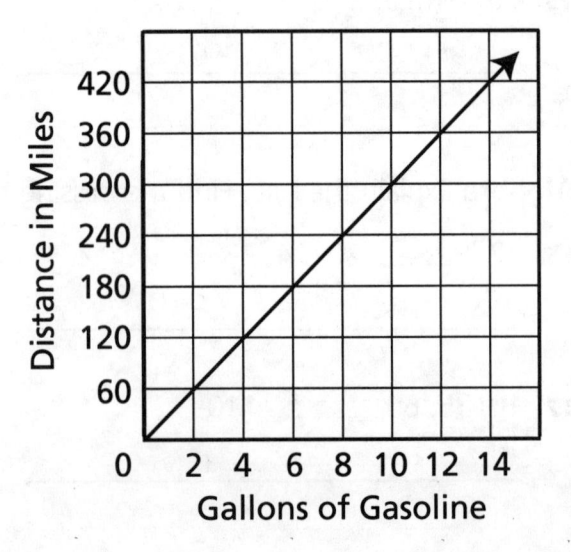

_____

78

Use with Grade 6, Chapter 15, Lesson 4, pages 348–350.

Name_____

Linear and Nonlinear Functions

P 15-5 PRACTICE

Identify each function as linear or nonlinear.
Explain your answer.

1.

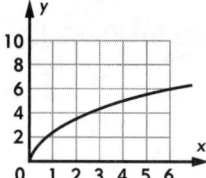

2.

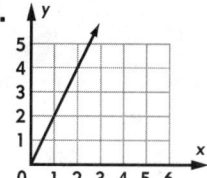

_____   _____
_____   _____

3.

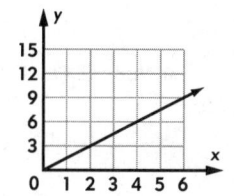

4.

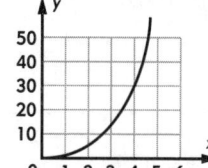

_____   _____
_____   _____

5.

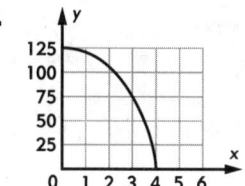

6.

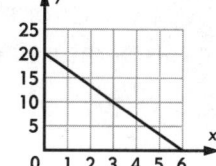

_____   _____
_____   _____

Graph the function. Identify it as linear or nonlinear. Explain.

7. $y = 2x^2$　　　　　**8.** $y = 2x - 1$　　　　　**9.** $y = x^3 + 1$

_____　　　_____　　　_____
_____　　　_____　　　_____
_____　　　_____　　　_____
_____　　　_____　　　_____

Use with Grade 6, Chapter 15, Lesson 5, pages 352–353.

Name _____

Problem Solving: Skill
Describe Relationships

P 15-6 PRACTICE

Use data from the graph for problems 1–3.

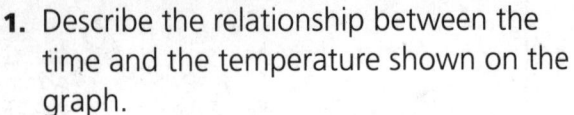

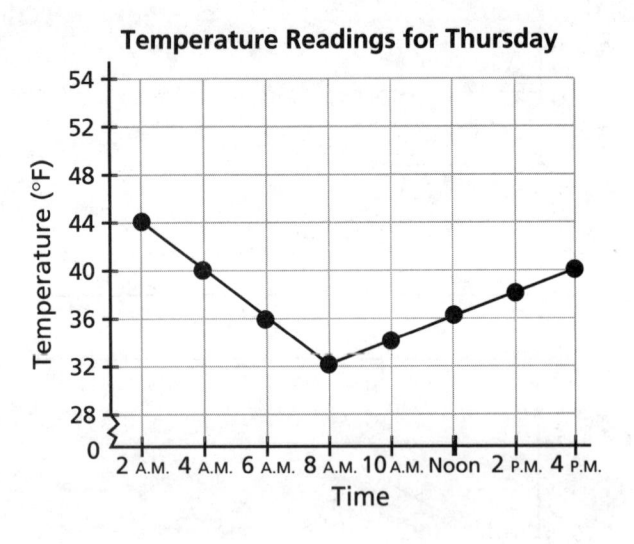

1. Describe the relationship between the time and the temperature shown on the graph.

   _____
   _____
   _____
   _____

2. What was the temperature at 2 P.M.?

   _____

3. Is the function linear? Explain.

   _____

Temperature Readings for Thursday

Choose the correct answer.

The air temperature increases by 1°C for every 160 meters you descend from the top of a mountain. The temperature at the top of the mountain is 18°C.

4. Which statement is true?

   A The temperature increases as you ascend to the top of the mountain.

   B The temperature remains steady after descending 150 meters.

   C The temperature increases as you descend the mountain.

   D The temperature at the bottom of the mountain is 18°C.

Mixed Strategy Review

5. A rectangular yard is 20 meters by 28 meters. How many fence posts would be required to put a fence around the yard if the posts are placed 4 meters apart?

   _____

6. An artist mixes pairs of different colors using the primary colors red, blue, and yellow. How many possible colors can he produce? Assume the artist uses equal amounts for each mixture.

   _____

80

Use with Grade 6, Chapter 15, Lesson 6, pages 354–355.

Name_____

Explore Addition Equations

P 16-1 PRACTICE

Solve the equation shown in each model.
There are the same number of algebra tiles on both sides of the mat.

Each ▮ represents the hidden number of tiles, x.

1. $x + 1 = 4$

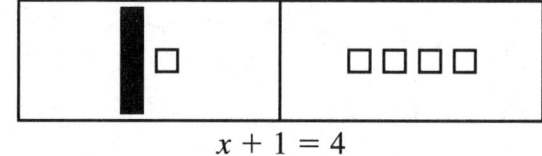

2. $x + 4 = 6$

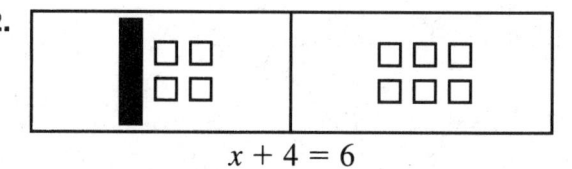

3. $3 = x + 1$

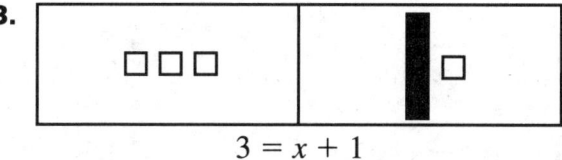

4. $x + 3 = 4$

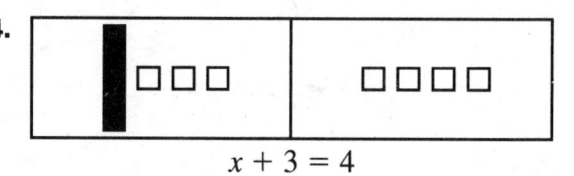

5. $x + 2 = 2$

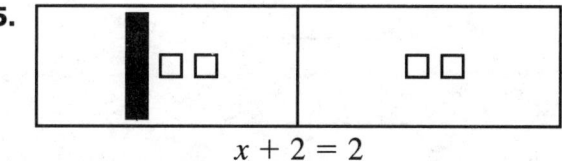

6. $x + 2 = 10$

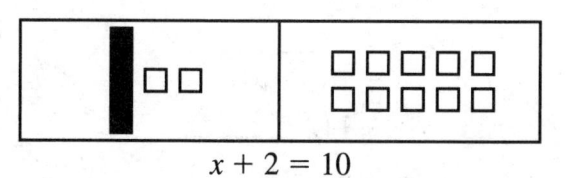

Solve.

7. $33 = k + 7$ _____

8. $g + 8 = 84$ _____

9. $62 + r = 83$ _____

10. $6 + w = 9$ _____

11. $x + 3 = 9$ _____

12. $42 + h = 52$ _____

13. $d + 28 = 45$ _____

14. $59 + d = 72$ _____

15. $g + 9 = 45$ _____

16. $x + 19 = 19$ _____

17. $t + 19 = 47$ _____

18. $14 + v = 76$ _____

19. $49 = c + 19$ _____

20. $b + 24 = 52$ _____

21. $16 = j + 5$ _____

Use with Grade 6, Chapter 16, Lesson 1, pages 360–361.

81

Name _____

Addition and Subtraction Equations

P PRACTICE 16-2

Solve. Check the solution.

1. $a - 48 = 563$ _____
2. $z + 18.9 = 26.3$ _____
3. $v - 4 = 8.76$ _____
4. $b - 2\frac{3}{4} = 1\frac{1}{4}$ _____
5. $m + 76 = 333$ _____
6. $7 + y = 7$ _____
7. $x + 23\frac{1}{3} = 354$ _____
8. $354 + x = 423\frac{1}{3}$ _____
9. $b + 18 = 46.9$ _____
10. $t + 18.7 = 53$ _____
11. $65.13 + r = 89.7$ _____
12. $z + 0 = 0$ _____
13. $v - 8.2 = 16.4$ _____
14. $424 = s + 86$ _____
15. $n - 5 = 5$ _____
16. $39\frac{2}{3} + t = 290$ _____
17. $5.555 + n = 15$ _____
18. $x + 0.4 = 5.1$ _____
19. $27 = e + 15$ _____
20. $w - 38 = 45$ _____
21. $125 = b + 112$ _____
22. $98 = t - 7$ _____
23. $32 + y = 83$ _____
24. $26 = 8 + v$ _____

Solve. Explain your solution.

25. Yesterday Jane mailed some cards. Today she mailed eight more cards. If Jane mailed 52 cards in all, how many did she mail yesterday?

    _____
    _____

26. Yesterday Josh sold some boxes of greeting cards. Today he sold seven boxes. If he sold 25 boxes in all, how many did he sell yesterday?

    _____
    _____

Name_____

Multiplication and Division Equations

P 16-3 PRACTICE

Solve. Check the solution.

1. $4h = 112$ _____
2. $v \div 47 = 96$ _____
3. $43n = 301$ _____
4. $18f = 504$ _____
5. $\frac{t}{21} = 21$ _____
6. $\frac{a}{63} = 5{,}607$ _____
7. $30b = 1{,}308$ _____
8. $\frac{m}{48} = 11.2$ _____
9. $\frac{918}{j} = 18$ _____
10. $\frac{w}{9} = 477$ _____
11. $5.5k = 0$ _____
12. $\frac{4}{324}z = \frac{4}{324}$ _____
13. $103m = 10{,}609$ _____
14. $\frac{1{,}000}{n} = 250$ _____
15. $\frac{h}{28} = 34$ _____
16. $8c = 4{,}896$ _____
17. $\frac{t}{12} = 12$ _____
18. $0.125z = 80$ _____

Solve. Use mental math, paper and pencil, or a calculator.

19. $3k = 66$ _____
20. $12a = 48$ _____
21. $w \div 25 = 4$ _____
22. $\frac{d}{9} = 16$ _____
23. If y is 7, what is $21y$?
24. If $8f = 72$, what is f?

_____   _____

Problem Solving

Solve. Explain your solution.

25. Skylar bought seven books for $87.50. How much did each book cost?

26. Clarinda has to make 96 cookies for school. How many dozens of cookies is this?

Use with Grade 6, Chapter 16, Lesson 3, pages 366–369.

Two-Step Equations

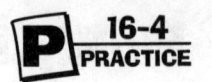

Solve. Check the solution.

1. $7m + 8 = 71$ _____
2. $\frac{y}{7} + 6 = 11$ _____
3. $12y + 2 = 146$ _____
4. $\frac{y}{12} + 1 = 6$ _____
5. $2a - 1 = 19$ _____
6. $\frac{c}{9} - 8 = 17$ _____
7. $4t + 16 = 24$ _____
8. $4f - 11 = 29$ _____
9. $\frac{g}{17} + 8 = 10$ _____
10. $13a - 9 = 17$ _____
11. $5c - 42 = 73$ _____
12. $4w - 26 = 82$ _____
13. $7a + 4 = 46$ _____
14. $\frac{m}{6} + 8 = 12$ _____
15. $9n - 7 = 56$ _____
16. $3k + 6 = 30$ _____
17. $\frac{x}{8} - 3 = 3$ _____
18. $15y - 40 = 35$ _____

Complete each table.

19.

| Rule: $y = 0.2x + 4$ ||
x	y
9	
40	
	8
	4.1

20.

| Rule: $z = \frac{w}{4} - 2$ ||
w	z
8	
	23
10	
	3

Problem Solving

Solve. Explain your solution.

21. Five more than half a number equals fifteen. What is the number?

22. Two less than three times a number equals thirteen. What is the number?

Name _____

Problem Solving: Strategy
Write an Equation

P 16-5 PRACTICE

Write an equation. Solve for the value given.

1. The number of fifth-grade students who won awards in the science fair last year was 3 times as many as this year less 2. If 16 fifth graders won awards last year, how many won awards this year?

   _____

2. The number of students who sing in the school chorus this year is twice the number as last year plus 12. If 96 students sing in the chorus this year, how many sang in the chorus last year?

   _____

3. Northside Soccer League has twice the number of members as Southside Soccer League less 130. If Northside Soccer League has 240 members, how many members does Southside Soccer League have?

   _____

4. **Health** The number of students at Albright Middle School who received fitness awards this year was 8 less than $\frac{1}{3}$ the number who received the fitness award last year. If 24 students received the fitness award this year, how many received it last year?

   _____

Mixed Strategy Review
Solve. Use any strategy.

5. Percy has a $20 bill, a $5 bill, and a $1 bill. How many different total amounts of money can he make with these bils?

   _____

   Strategy: _____

6. What are three consecutive even numbers that have a product of 480?

   _____

   Strategy: _____

7. A rectangular table seats 10 people: 4 on each side and one on each end. For a banquet, Ms. Mitchell wants to line up the tables end-to-end to form one long table that will seat 50 people. How many rectangular tables will she need?

   _____

   Strategy: _____

8. **Write a problem** for which you could write an equation to solve. Share it with others.

   _____

   _____

Use with Grade 6, Chapter 16, Lesson 5, pages 374–375.

Name _____

Use Properties of Operations

P 16-6 PRACTICE

Identify all the terms and name the coefficient of *w* in each expression.

1. $8w + 2m$ _____
2. $w + 6$ _____
3. $16 + 4w$ _____
4. $9w - 7$ _____
5. $7w - y$ _____
6. $3x + 5w$ _____
7. $10w - 6x$ _____
8. $2 + 21w$ _____

Simplify. Tell if the expression is a monomial.

9. $9x + 2x$ _____
10. $5y + 3z$ _____
11. $6w + 8w$ _____
12. $18k - 3k$ _____
13. $2a + 6 + a$ _____
14. $9b - 2b$ _____
15. $6x - 5x + 1$ _____
16. $14s + 17s$ _____

Solve each equation. Check the solution.

17. $2x + 4x = 24$ _____
18. $5y + 9y = 28$ _____
19. $3m - 2m = 8$ _____
20. $9k - 5k = 12$ _____
21. $5a - a = 40$ _____
22. $6b + 2b = 56$ _____

Identify the properties.

23. $9 \times 28 = (9 \times 20) + (9 \times 8)$

24. $(a + b) + c = a + (b + c)$

25. $a \times b = b \times a$

26. $6 \times 29 = (6 \times 30) - (6 \times 1)$

86

Use with Grade 6, Chapter 16, Lesson 6, pages 376–379.

Name_____

Explore Inequalities

P 16-7 PRACTICE

Solve. Graph the solution on the number line.

1. $x + 5 > 8$

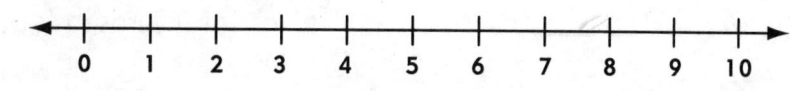

2. $w - 1 < 3$

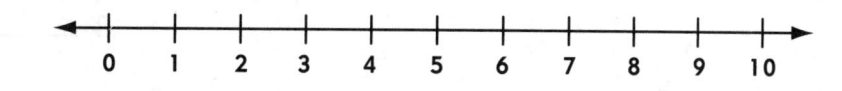

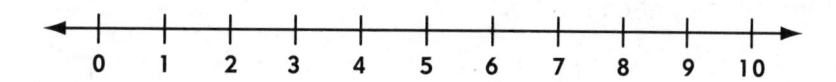

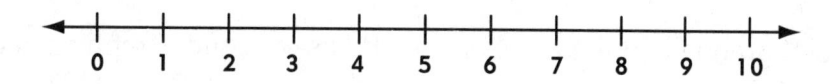

3. $c + 3 \geq 8$

4. $y + 1 \leq 7$

5. $m - 6 > 2$

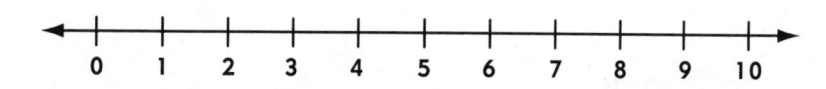

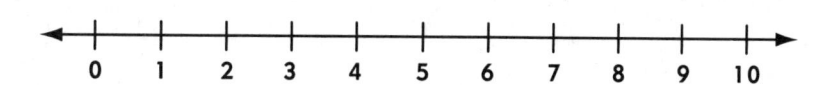

6. $b + 9 \geq 12$

7. $s - 3 < 3$

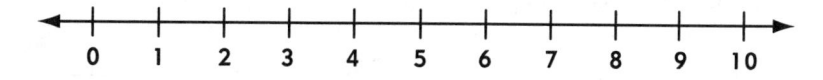

Use with Grade 6, Chapter 16, Lesson 7, pages 380–381.

Name _____

Integers and the Number Line

P 17-1 PRACTICE

Write an integer to represent each situation.

1. 20 feet below sea level _____
2. loss of 9 yards in football _____
3. 150 feet above sea level _____
4. savings of $94 _____

Find the opposite of each integer.

5. ⁻7 _____
6. 5 _____
7. ⁻8 _____
8. 16 _____
9. 4 _____
10. 11 _____

Find the absolute value of each integer.

11. ⁻6 _____
12. 3 _____
13. ⁻10 _____
14. ⁻15 _____
15. ⁻23 _____
16. 19 _____

Name each point graphed on the number line and its opposite. Graph each opposite integer.

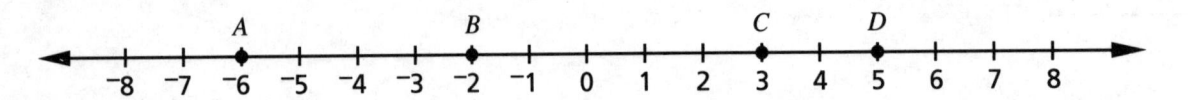

```
         A           B           C   D
<--+--+--●--+--+--+--●--+--+--+--+--●--●--+--+--+-->
  ⁻8 ⁻7 ⁻6 ⁻5 ⁻4 ⁻3 ⁻2 ⁻1  0  1  2  3  4  5  6  7  8
```

17. A _____
18. B _____
19. C _____
20. D _____

Compare. Write > or <.

21. ⁻8 ◯ ⁻2
22. 9 ◯ ⁻121
23. ⁻5 ◯ 0
24. ⁻11 ◯ ⁻21
25. ⁻4 ◯ 1
26. ⁻16 ◯ ⁻49

Write the integers in order from least to greatest.

27. ⁻3, ⁻2, 0 _____
28. ⁻3, 4, ⁻5 _____
29. 12, ⁻8, ⁻5 _____
30. 14, ⁻17, ⁻13, 11 _____
31. ⁻33, 32, ⁻36, 31 _____
32. ⁻59, 65, 43, ⁻86, ⁻94 _____

Problem Solving
Solve.

33. Joe lost $156 investing in stocks. How is this written as an integer?

34. A rare fish was found 35 feet below sea level. How is this written as an integer?

Name _____

Explore Adding Integers

P 17-2 PRACTICE

Complete the number sentence for each model.

1. ⊖ ⊖ ⊖
   ⊕ ⊕ ⊕ ⊕

   ⁻3 + 4 = _____

2. ⊖ ⊖ ⊖ ⊖ ⊖ ⊖
   ⊕ ⊕ ⊕ ⊕ ⊕ ⊕

   ⁻6 + 6 = _____

3. ⊕ ⊕ ⊕ ⊕ ⊕ ⊕
   ⊖ ⊖ ⊖ ⊖

   6 + ⁻4 = _____

4. ⊖ ⊖ ⊖
   ⊕ ⊕ ⊕ ⊕ ⊕

   ⁻3 + 5 = _____

5. ⊕ ⊕ ⊕
   ⊖ ⊖ ⊖ ⊖ ⊖

   3 + ⁻5 = _____

6. ⊕ ⊕
   ⊖ ⊖ ⊖ ⊖

   2 + ⁻4 = _____

Add. You may use counters.

7. 5 + ⁻6 = _____
8. 0 + ⁻4 = _____
9. ⁻3 + 7 = _____
10. ⁻8 + ⁻2 = _____
11. 5 + 18 = _____
12. ⁻7 + 0 = _____
13. ⁻2 + 8 = _____
14. ⁻16 + 16 = _____
15. ⁻2 + ⁻6 = _____
16. 0 + ⁻9 = _____
17. ⁻9 + ⁻5 = _____
18. ⁻11 + 4 = _____
19. ⁻9 + 8 = _____
20. ⁻15 + 6 = _____
21. ⁻6 + ⁻2 = _____
22. 12 + ⁻3 = _____
23. ⁻7 + ⁻7 = _____
24. ⁻20 + 20 = _____
25. ⁻12 + 0 = _____
26. 3 + ⁻10 = _____
27. ⁻9 + ⁻8 = _____

Problem Solving
Solve.

28. A football team gained 6 yd on one play. On the next play, the team lost 11 yd. Write the total gain or loss of yards as an integer.

    _____

29. At 7:30 A.M. the temperature was ⁻4°F. By 11:32 A.M., the temperature had risen 49 degrees. What was the temperature then?

    _____

Use with Grade 6, Chapter 17, Lesson 2, pages 400–401.

Add Integers

17-3 PRACTICE

Add.

1. ⁻12 + 5 = _____
2. ⁻7 + ⁻5 = _____
3. 10 + 6 = _____
4. ⁻15 + 15 = _____
5. 11 + ⁻13 = _____
6. ⁻10 + 2 = _____
7. 17 + ⁻19 = _____
8. ⁻20 + 4 = _____
9. 10 + ⁻11 = _____
10. ⁻4 + 16 = _____
11. 7 + ⁻14 = _____
12. ⁻14 + 8 = _____
13. 30 + ⁻8 = _____
14. ⁻12 + 2 = _____
15. 13 + ⁻7 = _____
16. ⁻21 + 12 = _____
17. 4 + ⁻4 = _____
18. 7 + ⁻8 = _____
19. 1 + ⁻7 = _____
20. 3 + ⁻6 = _____
21. ⁻2 + ⁻3 = _____

Find the sum. Identify the reason for each step.

22.   27 + (43 + ⁻27)
    = (27 + 43) + ⁻27 ⟵ _____
    = (43 + 27) + ⁻27 ⟵ _____
    = 43 + (27 + ⁻27) ⟵ _____
    = 43 + 0 ⟵ _____
    = 43 ⟵ _____

Name the addition rule used for each function table. Complete the table.

23.
Rule:					
Input	⁻19	⁻50	28	19	3
Output	⁻24	⁻55	23	14	

24.
Rule:					
Input	⁻7	⁻12	35	⁻2	⁻18
Output	⁻4	⁻9	38	1	

Problem Solving

Solve.

25. Steve is standing at sea level. He walks 9 ft down, then 4 ft up, then 3 ft down a tunnel. How many feet above or below sea level is he standing now? Write the answer as an integer.

26. The temperature at 8 A.M. was ⁻5°C. At 10 A.M. the temperature was 3°C warmer. At 4 P.M. the temperature was 4°C colder than at 10 A.M. What was the temperature at 4 P.M.?

Name _____

Explore Subtracting Integers

P 17-4 PRACTICE

Complete the number sentence for each model.

1. 4 − 3 = _____

2. ⁻4 − ⁻2 = _____

3. ⁻5 − ⁻1 = _____

4. 3 − ⁻1 = _____

Subtract. You may use counters.

5. 2 − 5 = _____   6. ⁻6 − ⁻7 = _____   7. ⁻2 − 4 = _____

8. 8 − ⁻5 = _____   9. 5 − 9 = _____   10. ⁻5 − 5 = _____

11. 5 − ⁻6 = _____   12. 7 − 4 = _____   13. ⁻8 − ⁻2 = _____

14. 9 − 3 = _____   15. ⁻10 − ⁻3 = _____   16. 4 − 9 = _____

17. ⁻16 − ⁻3 = _____ 18. ⁻10 − ⁻7 = _____   19. ⁻6 − 12 = _____

20. 3 − 5 = _____   21. ⁻18 − 5 = _____   22. 2 − 6 = _____

23. 3 − ⁻12 = _____   24. ⁻4 − 10 = _____   25. 2 − ⁻6 = _____

26. ⁻8 − ⁻14 = _____ 27. 12 − ⁻12 = _____   28. ⁻12 − 12 = _____

29. ⁻16 − ⁻16 = _____ 30. 4 − ⁻6 = _____   31. 8 − 16 = _____

Problem Solving
Solve.

32. On Monday, the highest temperature recorded was 11°F and the lowest temperature recorded was ⁻12°F. What is the difference in temperature between the highest and the lowest temperatures?

33. The temperature was 32°F on Tuesday. If the temperature dropped 40°F by Thursday, what was the temperature on Thursday?

Use with Grade 6, Chapter 17, Lesson 4, pages 406–407.

Name _____

Subtract Integers

PRACTICE 17-5

Subtract.

1. 3 − ⁻6 = _____
2. ⁻3 − ⁻6 = _____
3. ⁻8 − 2 = _____
4. ⁻7 − 4 = _____
5. ⁻5 − 6 = _____
6. 15 − ⁻4 = _____
7. 3 − ⁻12 = _____
8. 5 − ⁻5 = _____
9. 14 − 16 = _____
10. ⁻12 − 6 = _____
11. 2 − ⁻4 = _____
12. 8 − 3 = _____
13. 8 − ⁻2 = _____
14. ⁻10 − 2 = _____
15. 9 − ⁻3 = _____
16. ⁻12 − 3 = _____
17. ⁻2 − 10 = _____
18. ⁻18 − 10 = _____
19. 3 − 15 = _____
20. 15 − ⁻9 = _____
21. ⁻1 − 2 = _____
22. 4 − ⁻4 = _____
23. ⁻8 − ⁻2 = _____
24. ⁻5 − ⁻5 = _____
25. ⁻11 − ⁻6 = _____
26. 21 − ⁻7 = _____
27. ⁻13 − 13 = _____
28. ⁻7 − ⁻13 = _____
29. ⁻6 − 9 = _____
30. 7 − 19 = _____

Evaluate.

31. ⁻4 − (⁻2 + 5) = _____
32. 8 − ⁻3 + ⁻6 = _____
33. ⁻1 − ⁻9 − ⁻4 = _____
34. ⁻7 − ⁻2 − ⁻5 = _____
35. ⁻8 − ⁻3 − 5 = _____
36. ⁻12 − (⁻6 − 1) = _____
37. ⁻16 − ⁻9 + 2 = _____
38. 15 − ⁻4 + ⁻3 = _____
39. 10 − ⁻3 − ⁻5 = _____
40. ⁻11 + ⁻1 − ⁻6 = _____

Problem Solving
Solve.

41. After a rocket reached an altitude of 13,480 ft, it separated from the main engines. The engines sank into the ocean to a depth of ⁻1,550 ft. How far did the engines fall?

42. When an airplane flew at an altitude of 5,000 ft, the temperature was ⁻15°F outside. When the airplane reached an altitude of 10,000 ft, the temperature was ⁻28°F outside. What was the difference in temperature?

Name _____

Problem Solving: Skill
Check for Reasonableness

17-6 PRACTICE

Find whether each answer is reasonable. Solve.

1. At 3 A.M., the tide was ⁻4 feet. By 9 A.M., the tide had risen 6 feet. Andre calculates that the tide reached 10 feet at 9 A.M. Is his calculation reasonable? Explain.

2. The low temperature on Monday is ⁻5°C. The low temperature the following Monday is ⁻1°C. Gretchen calculates that the range in the temperatures is 6°C. Is her calculation reasonable? Explain.

3. A scuba diver descended 8 feet below the surface of the water. Then he descended an additional 12 feet. He calculates that he is ⁻20 feet from the surface of the water. Is his calculation reasonable? Explain.

4. A hiker descends 15 feet into a canyon. Then she descends another 9 feet on her hike. She calculates that her elevation is ⁻6 feet from where she began her descent. Is her calculation reasonable? Explain.

Choose the correct answer.

At Clearview Beach, the lowest tide of the year was ⁻9 feet. The highest tide for the year was 12 feet.

5. Which of the following is true?

   A The range in the highest and lowest tides for the year is 21 feet.
   B The tide changes by 3 feet from high tide to low tide.
   C The highest tide for the year reached 21 feet.
   D The lowest tide for the year reached ⁻12 feet.

6. When checking if an answer is reasonable,

   F rework the problem at least three times.
   G compare it with known facts.
   H guess whether or not the answer looks correct.
   J use multiplication to solve.

Mixed Strategy Review

7. A croquet ball has a mass of 460 grams. Together, the mass of a golf ball and a croquet ball is the same as the mass of 11 golf balls. What is the mass of one golf ball?

8. The temperature recorded at 5 A.M. was 25°F. The temperature increased by 2°F for every hour for the next four hours. What was the temperature at the end of the four hours?

Use with Grade 6, Chapter 17, Lesson 6, pages 412–413.

Name _____

Multiply and Divide Integers

P 17-7 PRACTICE

Multiply or divide. Multiply to check division.

1. 8 × ⁻3 = _____
2. ⁻3 × ⁻10 = _____
3. 10 × ⁻4 = _____

4. 2 × ⁻5 = _____
5. 9 × ⁻9 = _____
6. ⁻9 × ⁻7 = _____

7. 8 × ⁻6 = _____
8. ⁻7 × 8 = _____
9. 7 × ⁻3 = _____

10. ⁻13 × 1 = _____
11. ⁻3 × ⁻6 = _____
12. ⁻2 × ⁻2 = _____

13. ⁻3 × ⁻13 = _____
14. 10 × ⁻5 = _____
15. 5 × ⁻4 = _____

16. ⁻4 × ⁻8 = _____
17. 9 × ⁻4 = _____
18. ⁻4 × ⁻11 = _____

19. ⁻84 ÷ 12 = _____
20. 63 ÷ ⁻7 = _____
21. $\frac{81}{-3}$ = _____

22. ⁻108 ÷ 12 = _____
23. ⁻17 ÷ 17 = _____
24. $\frac{-76}{4}$ = _____

25. 15 ÷ ⁻5 = _____
26. ⁻19 ÷ ⁻1 = _____
27. $\frac{108}{-9}$ = _____

28. $\frac{-120}{-20}$ = _____
29. 80 ÷ ⁻16 = _____
30. ⁻99 ÷ ⁻33 = _____

31. $\frac{-27}{3}$ = _____
32. 14 ÷ 7 = _____
33. ⁻24 ÷ 8 = _____

34. 56 ÷ ⁻7 = _____
35. $\frac{-70}{10}$ = _____
36. 42 ÷ 7 = _____

Evaluate each expression.

37. 18 ÷ ⁻2 × 8 _____
38. 28 ÷ ⁻4 × 6 _____

39. 36 ÷ (⁻3 × 3) _____
40. 13 + 18 ÷ ⁻6 _____

41. 10 × ⁻4 ÷ 5 − ⁻6 _____
42. 16 − (25 ÷ ⁻5) _____

Problem Solving
Solve.

43. Which has the greatest quotient: 25 ÷ ⁻5; ⁻25 ÷ ⁻5; ⁻25 ÷ 5; or 25 ÷ 5? Explain.

44. The price of stock in the Omega Company for Monday went up $5 per share. If Judy owns 32 shares, how much did her stock holdings change in value? Write this as an integer.

94

Use with Grade 6, Chapter 17, Lesson 7, pages 414–416.

Name _____

Negative Exponents

P 18-1 PRACTICE

Evaluate.

1. 9^{-2} _____
2. 5^{-4} _____
3. 11^{-1} _____

4. 4^{-2} _____
5. 13^{-2} _____
6. 3^{-4} _____

7. 6^{-4} _____
8. 8^{-3} _____
9. 7^{-2} _____

10. 10^{-1} _____
11. 9^{-3} _____
12. 3^{-6} _____

13. 11^{-2} _____
14. 12^{-2} _____
15. 15^{-1} _____

16. 465^{0} _____
17. 16^{0} _____
18. 21^{-2} _____

Write each expression as a fraction with a positive exponent.

19. 9^{-12} _____
20. 7^{-48} _____
21. 25^{-9} _____

22. 38^{-58} _____
23. 21^{-104} _____
24. 6^{-987} _____

25. 8^{-5} _____
26. 10^{-10} _____
27. 16^{-7} _____

28. 3^{-31} _____
29. 11^{-20} _____
30. 5^{-8} _____

Write each as an integer with a negative exponent.

31. $\dfrac{1}{4^{10}}$ _____
32. $\dfrac{1}{3^{6}}$ _____
33. $\dfrac{1}{6^{5}}$ _____

34. $\dfrac{1}{9^{11}}$ _____
35. $\dfrac{1}{7^{18}}$ _____
36. $\dfrac{1}{10^{10}}$ _____

37. $\dfrac{1}{15^{7}}$ _____
38. $\dfrac{1}{19^{14}}$ _____
39. $\dfrac{1}{31^{40}}$ _____

40. $\dfrac{1}{83^{25}}$ _____
41. $\dfrac{1}{55^{89}}$ _____
42. $\dfrac{1}{94^{107}}$ _____

Problem Solving

Solve.

43. The half-life of a radioactive form of calcium is 1 second. If you start with 180 grams, how much will be left after 3 seconds?

44. Suppose the half-life of a radioactive form of a substance is 1 hour. If you start with 1,600 grams, how much would be left after 4 hours?

Use with Grade 6, Chapter 18, Lesson 1, pages 422–423.

Name _____

Scientific Notation

P 18-2 PRACTICE

Evaluate.

1. $1.7 \times 10^4 =$ _____
2. $6.3 \times 10^{-4} =$ _____
3. $5.92 \times 10^6 =$ _____
4. $5.7 \times 10^{-3} =$ _____
5. $9 \times 10^{10} =$ _____
6. $2.8 \times 10^{-1} =$ _____
7. $1.45 \times 10^{-5} =$ _____
8. $3.04 \times 10^3 =$ _____
9. $8.65 \times 10^6 =$ _____
10. $9.3 \times 10^{-2} =$ _____
11. $2.68 \times 10^1 =$ _____
12. $1.23 \times 10^7 =$ _____

Write each number in scientific notation.

13. $0.009 =$ _____
14. $132,000 =$ _____
15. $91,000,000 =$ _____
16. $0.0015 =$ _____
17. $0.00008 =$ _____
18. $1,030,000 =$ _____
19. 76 million $=$ _____
20. $0.000034 =$ _____
21. $0.028 =$ _____
22. 59 billion $=$ _____
23. $720,000,000 =$ _____
24. $0.0000046 =$ _____
25. $0.00062 =$ _____
26. $6,020,000,000 =$ _____

Problem Solving
Solve.

27. The area of the Pacific Ocean is approximately 638,000,000 square miles. How is this written in scientific notation?

    _____

28. A virus has a diameter of 2×10^{-7} meters. How is this written in standard form?

    _____

Use with Grade 6, Chapter 18, Lesson 2, pages 424–425.

Name _____

Problem Solving: Strategy
Alternate Solution Methods

P 18-3 PRACTICE

Solve.

1. A scuba diver was 18 feet deep. He ascended 4 feet and then descended 6 feet. What was his depth then?

   _____

2. A submersible was 1,400 feet below sea level. It descended 300 feet. Then it rose 435 feet. Finally, it descended 280 feet. How far below sea level was it then?

   _____

3. The entrance to a cave is at sea level. A trail in the cave descends 23 feet. Then the trail ascends 6 feet to a large opening. How far below sea level is the large opening in the cave?

   _____

4. The temperature at 2 A.M. was ⁻4°F. By 6 A.M., the temperature had dropped 5°F. Then the temperature dropped another 2°F by 8 A.M. By 1 P.M., the temperature had risen 16°F. What was the temperature at 1 P.M.?

   _____

Mixed Strategy Review
Solve. Use any strategy.

5. Alan bought a car for $1,000. He planned to make some repairs and then sell it. He bought parts for the car for $228. Then he had to replace the transmission for $485. He was able to sell the car for $1,600. Did Alan make a profit or not? How much?

   _____

   Strategy: _____

6. Amanda is watching her neighbor's dog while her neighbor is on vacation. She will watch the dog for 2 weeks. If her neighbor is paying her $6 per day, how much money will Amanda earn?

   _____

   Strategy: _____

7. The Biltmores are buying mulch for their flower gardens. One garden is circular, with a diameter of 2 feet. The other garden is a rectangle, 3 feet by 4 feet. How many square feet of mulch do they need to buy to cover both gardens?

   _____

   Strategy: _____

8. **Write a problem** for which you could use an alternate solution method to solve. Share it with others.

   _____

   _____

Use with Grade 6, Chapter 18, Lesson 3, pages 426–427.

97

Name _____

Rational Numbers

P 18-4 PRACTICE

Show that each number is a rational number.

1. ⁻7 _____   2. $2\frac{1}{2}$ _____   3. ⁻3.03 _____

4. ⁻0.9 _____   5. $3\frac{7}{12}$ _____   6. ⁻$1\frac{5}{8}$ _____

7. ⁻15 _____   8. ⁻$2\frac{3}{12}$ _____   9. ⁻4.7 _____

10. ⁻$3\frac{1}{4}$ _____   11. 1.03 _____   12. ⁻0.79 _____

13. ⁻8 _____   14. 0.93 _____   15. 1.8 _____

16. ⁻$6\frac{1}{2}$ _____   17. 0.2 _____   18. $4\frac{1}{4}$ _____

Write the rational numbers as fractions with a common denominator.

19. ⁻0.5 and ⁻$\frac{9}{10}$ _____

20. 0.6 and ⁻$\frac{1}{2}$ _____

21. ⁻0.24 and $\frac{1}{25}$ _____

22. ⁻0.9 and ⁻1.43 _____

Write the rational numbers as decimals.

23. $2\frac{1}{2}$ and $3\frac{1}{2}$ _____

24. $1\frac{2}{5}$ and $1\frac{4}{5}$ _____

25. $2\frac{3}{100}$ and ⁻2.75 _____

26. ⁻3.45 and $2\frac{3}{4}$ _____

Compare. Write >, <, or =.

27. 0.9 ◯ ⁻0.7   28. ⁻4.5 ◯ ⁻$\frac{4}{3}$ 29. 3.1 ◯ $3\frac{3}{10}$

30. ⁻3.3 ◯ ⁻$3\frac{3}{10}$ 31. ⁻5.5 ◯ ⁻5.3   32. ⁻$6\frac{2}{3}$ ◯ ⁻6.6

Problem Solving
Solve.

33. The Dow Jones average had a high of 1,089.45 points on Monday. Express this rational number as a fraction.

_____

34. Ace Electronics had a record high of 24.5 for its stock. Express this rational number as a fraction.

_____

Use with Grade 6, Chapter 18, Lesson 4, pages 428–431.

Name _____

Operations on Rational Numbers

18-5 PRACTICE

Find the absolute value.

1. $^-9\frac{1}{2}$ _____
2. $\frac{5}{6}$ _____
3. $^-1.76$ _____
4. $^-5.132$ _____

Add, subtract, multiply, or divide.

5. $\frac{5}{6} + \frac{^-1}{2} =$ _____
6. $^-4\frac{1}{3} - {^-\frac{3}{5}} =$ _____
7. $\frac{1}{8} - \frac{^-1}{6} =$ _____
8. $^-2\frac{1}{3} - {^-1\frac{5}{12}} =$ _____
9. $^-10 - 3\frac{11}{12} =$ _____
10. $\frac{8}{9} \times \frac{^-3}{4} =$ _____
11. $\frac{^-2}{3} \times \frac{^-1}{8} =$ _____
12. $\frac{3}{4} \times \frac{^-2}{3} =$ _____
13. $^-3\frac{2}{5} \times 2\frac{1}{2} =$ _____
14. $^-6 \div \frac{3}{4} =$ _____
15. $^-3 \div {^-1\frac{1}{2}} =$ _____
16. $^-1\frac{3}{5} \div \frac{1}{4} =$ _____
17. $^-11.5 - {^-18.6} =$ _____
18. $^-12.2 \times {^-8.8} =$ _____
19. $^-6.7528 \div 1.84 =$ _____
20. $^-12.33 - 4.6 =$ _____
21. $39.7 \times {^-4.98} =$ _____
22. $^-9.612 \div {^-2.67} =$ _____
23. $^-4.8 \times 1.33 =$ _____
24. $^-6.67 + 89.33 =$ _____
25. $^-8.6 + {^-9.11} =$ _____
26. $^-4.62 \times {^-0.4} =$ _____

Algebra Name the rule used for each function table. Complete each table.

27.

Rule:					
Input	3.7	$^-2.9$	0	$^-6.1$	$^-3.4$
Output	$^-6.29$	4.93	0	10.37	

28.

Rule:					
Input	$^-3\frac{1}{4}$	$^-5\frac{1}{2}$	6	3	$4\frac{1}{2}$
Output	$^-5\frac{3}{4}$	$^-8$	$3\frac{1}{2}$	$\frac{1}{2}$	

Problem Solving

Solve.

29. On three sales, a stationery-store owner made $3.15 profit, a $4.85 loss, and an $0.50 profit. Find the net profit or loss.

30. A salmon swam upstream for 2 hours at a speed of $25\frac{1}{2}$ miles per hour. How far did the salmon swim?

_____ _____

Use with Grade 6, Chapter 18, Lesson 5, pages 432–435.

99

Graph Equations in Four Quadrants • Algebra

Practice 18-6

Name the coordinates of each point.

1. A _____
2. B _____
3. C _____
4. D _____
5. E _____
6. F _____

Name the point that each ordered pair represents.

7. (3, 0) _____
8. (⁻3, 0) _____
9. (⁻2, ⁻4) _____
10. (2, ⁻7) _____
11. (1, 4) _____
12. (⁻2, 5) _____

Complete each table. Then graph the function.

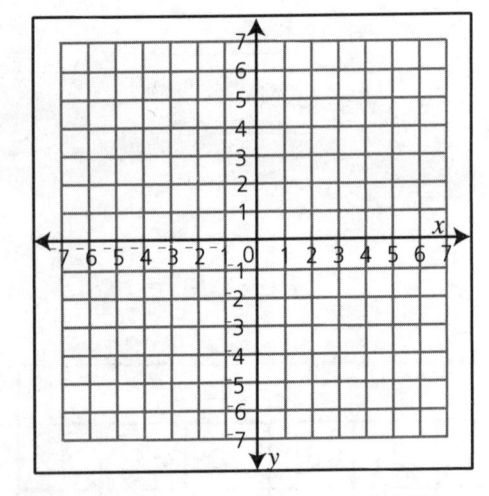

13. $y = x - 3$

x	⁻1	0	1	2
y				

14. $y = ⁻2x$

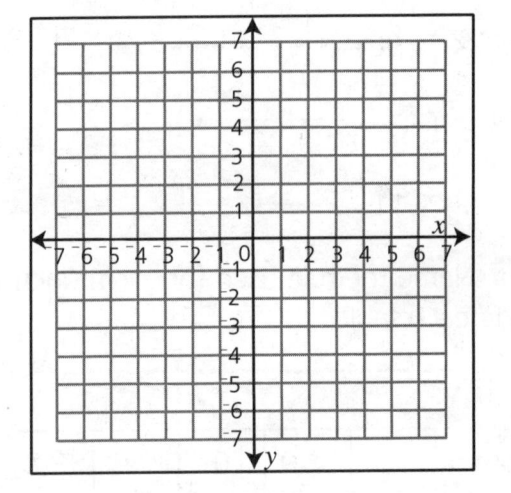

x	⁻1	0	1	2
y				

Problem Solving

Solve.

15. Ann uses the equation $y = x + 5$ to track a flock of migratory blue birds. Will the birds pass through the point (⁻1, ⁻4)? Explain.

_____

_____

16. An ant was located at the point (2, 3). It crawled down vertically 4 units, and turned to its right and crawled horizontally for 7 units. At what coordinates is the ant now?

_____

100

Use with Grade 6, Chapter 18, Lesson 6, pages 436–438.

Name _____

Perfect Squares and Square Roots

P 18-7 PRACTICE

Find the square and square root of each number.

1. 25 _____
2. 64 _____
3. 16 _____
4. 121 _____

5. 225 _____
6. 169 _____
7. 100 _____
8. 900 _____

Is the number a perfect square? If YES, find its square root.
If NO, find its square root, rounded to one decimal place.

9. 81 _____
10. 90 _____
11. 36 _____
12. 196 _____

13. 150 _____
14. 200 _____
15. 289 _____
16. 400 _____

17. 500 _____
18. 312 _____
19. 460 _____
20. 529 _____

21. 162 _____
22. 361 _____
23. 576 _____
24. 620 _____

25. 676 _____
26. 712 _____
27. 961 _____
28. 809 _____

Problem Solving
Solve.

29. If an object is dropped from a height of h feet, the time it takes to reach the ground is about $0.25 \times \sqrt{h}$ seconds. About how long would it take a ball dropped from a height of 256 feet to reach the ground?

30. Use the expression from problem 29. About how long would it take a ball dropped from a height of 400 feet to reach the ground?

Use with Grade 6, Chapter 18, Lesson 7, pages 440–441.

Name _____

Basic Geometry Terms

P 19-1 PRACTICE

Use a symbol to name each figure.

1.  •———————•→
   F   G

   _____

2.  ←—•—————————•→
     D   E

   _____

3.   •
        B

   _____

4.  •—————————•
   A   B

   _____

Draw a diagram of each figure and label it.

5. line RS

6. ray CD

7. $\overline{KL}$

8. point S

9. $\overline{NP}$

10. $\overrightarrow{QR}$

Name each polygon.

11. _____

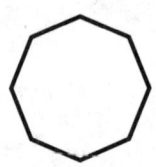

12. _____

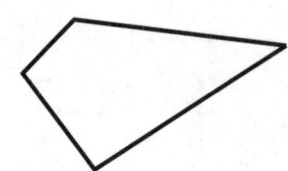

13. _____

14. _____

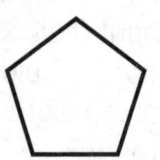

15. Draw an open figure with three line segments.

16. Draw a closed figure that is *not* a polygon.

Use with Grade 6, Chapter 19, Lesson 1, pages 456–458.

Name _____

Measure and Classify Angles

P 19-2
PRACTICE

Name each angle three ways. Then estimate, measure, and classify it.

1.

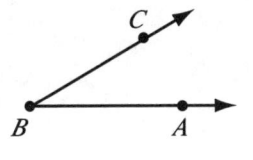

2.

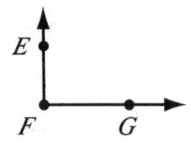

_____
_____

_____
_____

3.

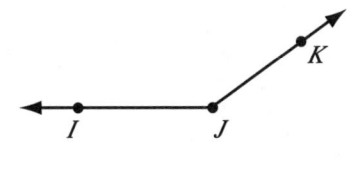

4.

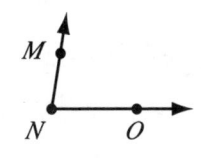

_____
_____

_____
_____

Draw an angle with each given measure.

5. 100°   6. 48°   7. 165°   8. 75°

Estimate the measure of each angle. Then measure it.

9.

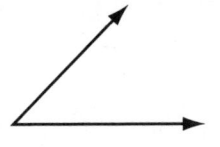

10.

_____

_____

11.

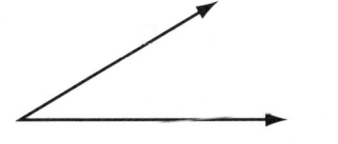

12.

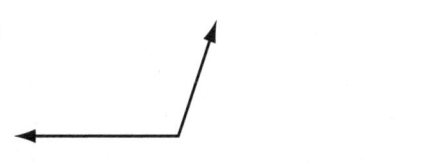

_____

_____

13.

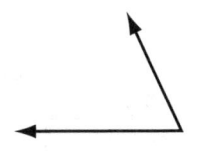

14.

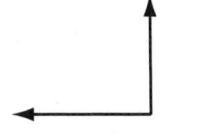

_____

_____

Use with Grade 6, Chapter 19, Lesson 2, pages 460–462.

Name_____

Lines and Angles

P 19-3 PRACTICE

Use the diagram for problems 1–7.

Use the diagram for problems 1–7.

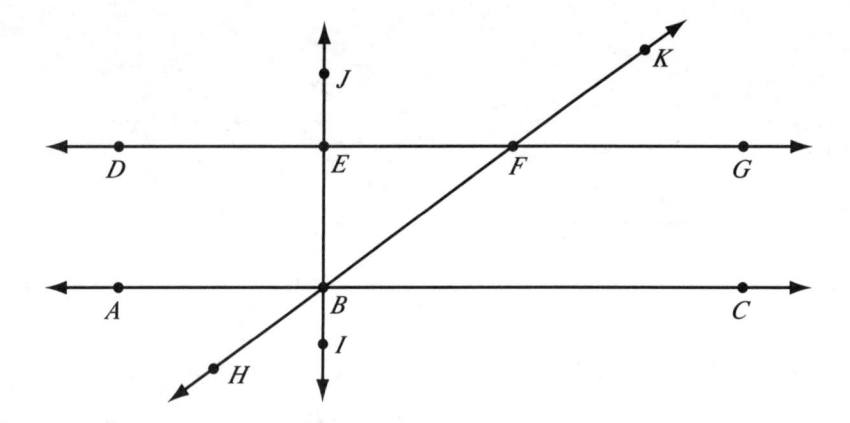

1. Name a pair of parallel lines.

_____

2. Name two pairs of perpendicular lines.

_____

3. Name a pair of adjacent angles.

_____
_____

4. Name two pairs of intersecting lines.

_____
_____

5. Name a pair of complementary angles.

_____
_____

6. Name a pair of supplementary angles.

_____
_____

7. Name a pair of alternate interior angles.

_____
_____

Find the unknown measure.

8. An angle measures 68°. What is the measure of a complementary angle?

_____

9. An angle measures 30°. What is the measure of a supplementary angle?

_____

10. An angle measures 16°. What is the measure of a complementary angle?

_____

11. An angle measures 167°. What is the measure of a supplementary angle?

_____

Name _____

Triangles

P 19-4 PRACTICE

Classify each triangle by its side lengths and angle measures.
If the triangle is a regular polygon, write *regular* as well.

1. 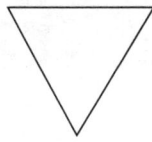 ▽   2. 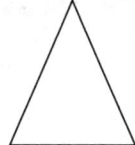 △   3. 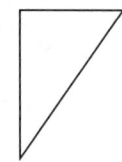 ◺

_____   _____   _____

Draw each triangle using a ruler. Use a protractor if you need to measure any angles.

4. acute triangle   5. right triangle

6. obtuse triangle   7. equilateral triangle

Algebra Find the measure of each unknown angle.

8. 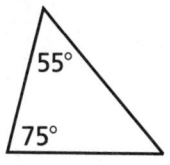 (55°, 75°) _____   9. 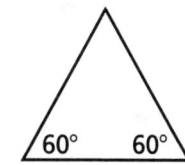 (60°, 60°) _____

10. 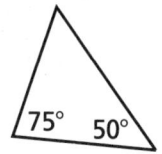 (75°, 50°) _____   11.  (44°, 68°) _____

12. 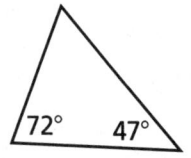 (72°, 47°) _____   13. 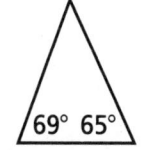 (69°, 65°) _____

Use with Grade 6, Chapter 19, Lesson 4, pages 468–471.

Name _____

Quadrilaterals

P 19-5 PRACTICE

Identify each quadrilateral.

1. _____

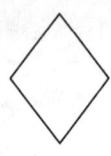

2. _____

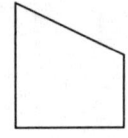

3. _____

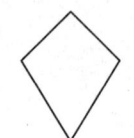

Draw each quadrilateral using a ruler.

4. rhombus

5. trapezoid

6. square

7. kite

8. parallelogram

9. rectangle

Algebra Find the measure of the unknown angle.

10. 55°, 110°, 140°, ?

_____

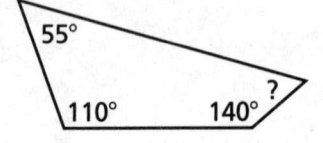

11. 90°, 150°, 90°, ?

_____

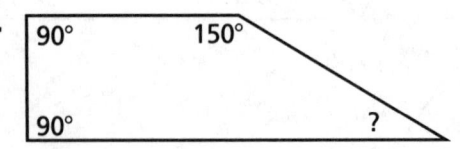

12. 110°, ?, 70°, 65°

_____

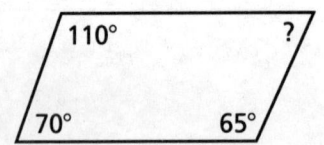

13. 90°, ?, 80°, 115°

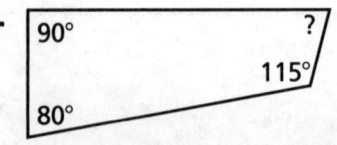

_____

106

Use with Grade 6, Chapter 19, Lesson 5, pages 472–475.

Name _____

Problem Solving: Skill
Use a Diagram

P 19-6 PRACTICE

Draw a diagram to solve.

1. Using any diagonals of your choice, show how you can divide a regular hexagon into 4 triangle sections and 2 rectangle sections.

2. Using any 4 lines of your choice, show how you can divide a square into 8 triangle sections.

3. How many diagonals that begin at the same vertex do you need to divide an octagon into 6 triangles?

   _____

   _____

   _____

4. Show how you can divide a pentagon into 2 quadrilateral sections and 1 triangle section using any 2 lines of your choice.

Choose the correct answer.

Alessandro drew a rectangle that was three times as long as it was wide. He divided it with two diagonals.

5. What number of sections and shapes did Alessandro create?

   A two square sections

   B three square sections

   C four triangle sections

   D six square sections

6. Which statement about diagrams is true? A diagram

   F should always have two sections.

   G makes information easier to see.

   H should always include a name.

   J makes it harder to solve any problem.

Mixed Strategy Review

7. The temperature rose 15°F in four hours. If the temperature is −2°F now, what was the temperature four hours ago?

   _____

8. Felicity has one $20-bill, one $10-bill, and one $5-bill. What possible totals can Felicity make with exactly two of the three bills?

   _____

Use with Grade 6, Chapter 19, Lesson 6, pages 476–477.

Circles

19-7 PRACTICE

Identify the parts of circle O.

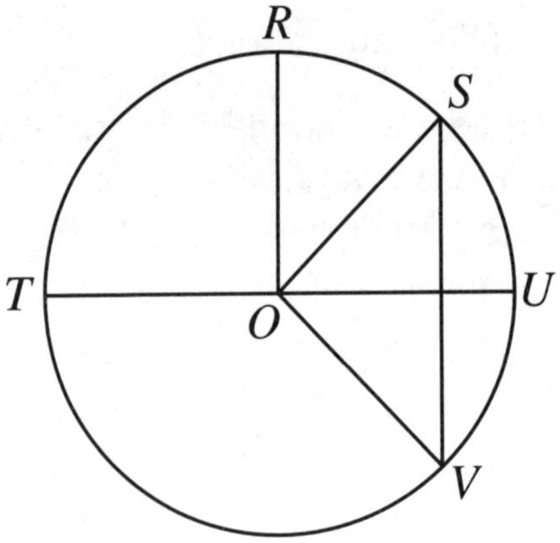

1. Name the radii. _____

2. Name the chords. _____

3. Name the diameters. _____

4. Name two arcs. _____

5. Classify triangle SOV by its sides _____

6. Name two central angles that have a measure less than 90°. _____

7. Name a central angle with a measure of 90°. _____

Draw a circle for each given radius on a separate piece of paper.

8. 24 mm

9. $2\frac{1}{2}$ in.

10. 3 cm

Draw a circle for each given diameter on a separate piece of paper.

11. 2 in.

12. 12 cm

13. 158 mm

Algebra Find the measure of each unknown central angle.

14. (circle with 64°, 106°, 132°)

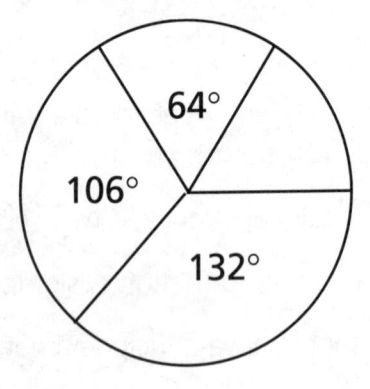

15. (circle with 68°, 73°, 89°, 35°)

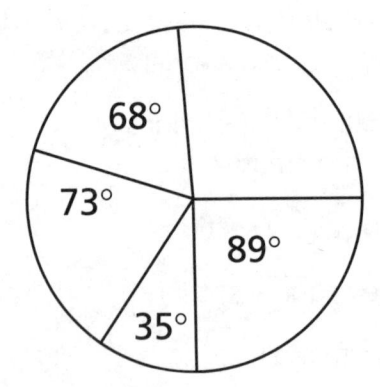

Problem Solving
Solve.

16. A circle has 10 central angles. All the central angles have the same measure. What is the measure of each central angle?

17. A circle has 4 central angles. Two of the angles are congruent. The third angle has a measure of 30° and the fourth angle has a measure of 68°. What is the measure of the other two central angles?

Name _____

Congruence and Similarity

P 20-1
PRACTICE

For each pair of figures, write congruent, similar but not congruent, or neither.

1.

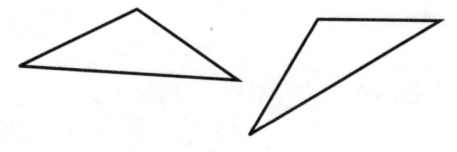

_____

2.

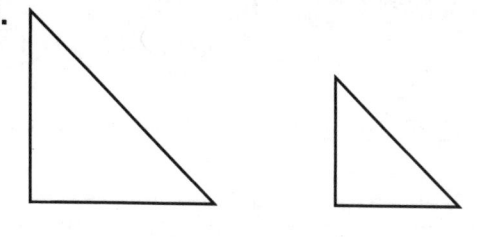

_____

3.

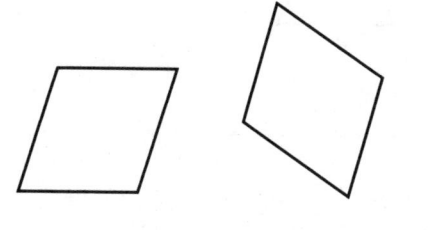

_____

4.

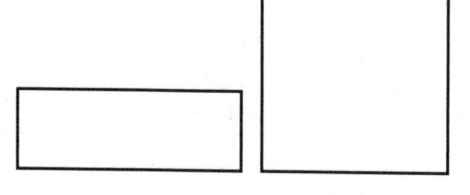

_____

5.

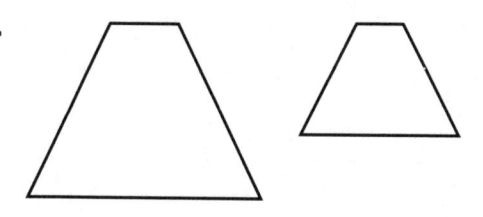

_____

6.

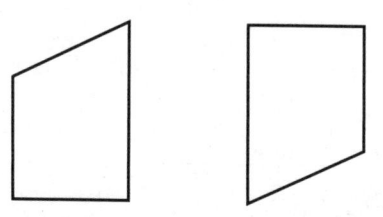

_____

7.

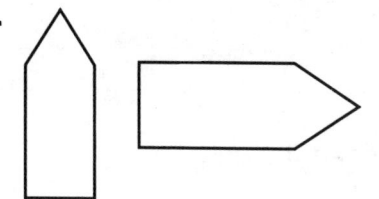

_____

8.

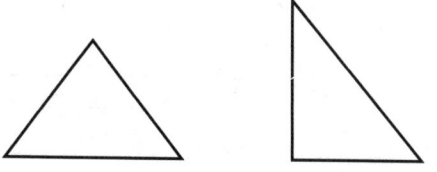

_____

Match each pair of congruent figures.

9.

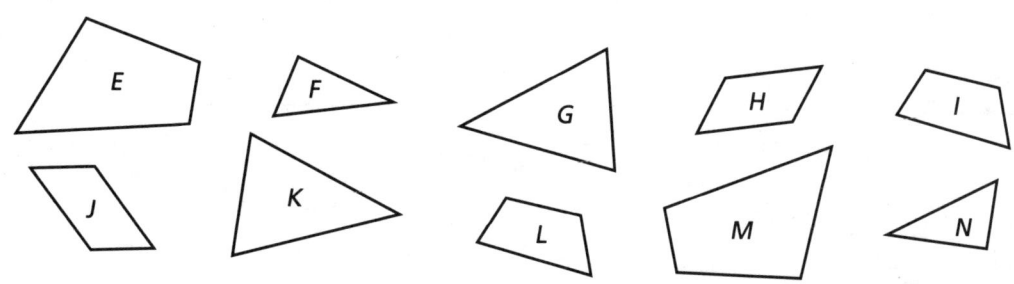

Use with Grade 6, Chapter 20, Lesson 1, pages 486–487.

109

Transformations

P 20-2 PRACTICE

Tell what transformation or combination of transformations changes figure A into congruent figure B.

1.

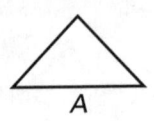

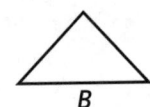

2.

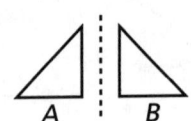

3.

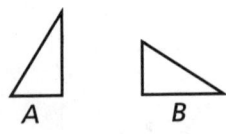

4.

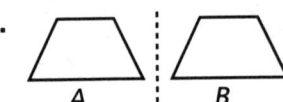

5.

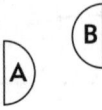

6.

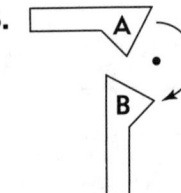

7.

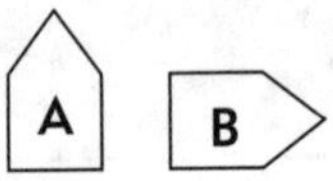

8.

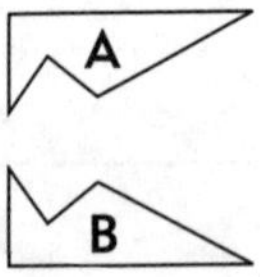

Tell what transformation or combination of transformations changed the original figure at the left into the image on the right.

9.

10.

11.

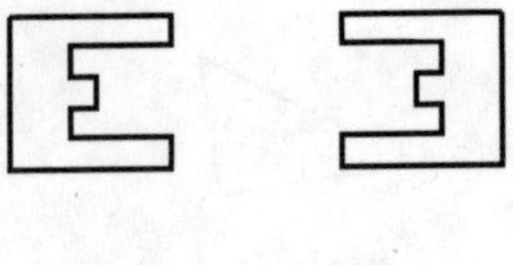

12.

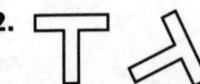

110

Use with Grade 6, Chapter 20, Lesson 2, pages 488–490.

Name _____

Symmetry

P 20-3 PRACTICE

Trace the figure. Does it have line symmetry?
Write *yes* or *no*. If *yes*, draw the lines of symmetry.

1.
_____

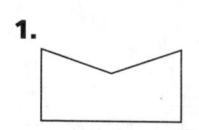

2.
_____

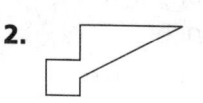

3.
_____

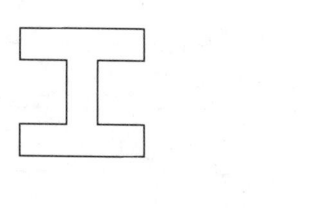

4.
_____

5.
_____

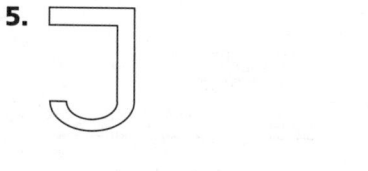

6.
_____

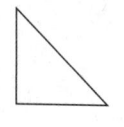

Tell whether the figure has rotational symmetry.
If it does, find the smallest fraction of one full turn
through which it can be turned and look the same.

7.
_____

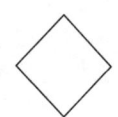

8.
_____

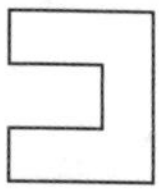

9.
_____

10.
_____

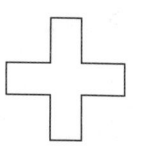

11.
_____

12.
_____

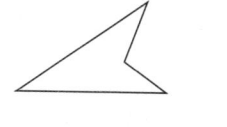

Use with Grade 6, Chapter 20, Lesson 3, pages 492–493.

Problem Solving: Strategy
Find a Pattern

Describe a pattern for each sequence.
Find the next shape in the sequence.

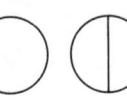

1.

2.

3.

4.

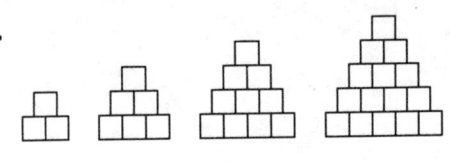

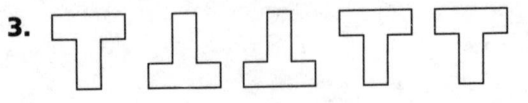

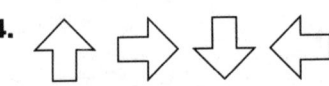

Mixed Strategy Review
Solve. Use any strategy.

5. Julie went for a bike ride. She left home and rode 2.5 miles east, 5 miles north, 3 miles west, and 8 miles south. In what direction and how many miles does she need to ride to get home?

   Strategy: _____

6. Leo's age is one third of Carrie's age. Carrie is 3 years younger than Samuel. Samuel is 27 years old. How old is Leo?

   Strategy: _____

7. **Write a problem** for which you could find a pattern to solve. Share it with others.

Name_____

Constructions

20-5 PRACTICE

Draw each line segment with a ruler. Draw each angle using a protractor. Then use a compass and ruler to draw the perpendicular bisector or angle bisector.

1. $\overline{ST}$ is 4 cm long.

2. $\overline{ST}$ is 50 mm long

3. $\overline{ST}$ is 2 in. long

4. $\overline{AB}$ is 6 cm long.

5. $m\angle A = 45°$

6. $m\angle D = 65°$

7. $m\angle F = 120°$

8. $m\angle E = 165°$

Use a compass and a straightedge to construct the perpendicular bisector or angle bisector.

9.

10.

11.

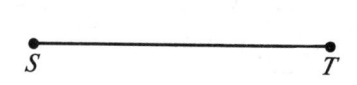

12.

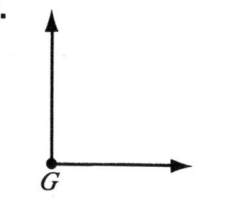

13.

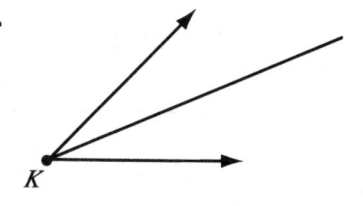

14.

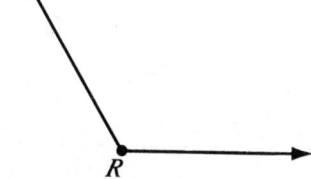

Use with Grade 6, Chapter 20, Lesson 5, pages 496–499.

113

Name_____

Explore Tessellations

P 20-6 PRACTICE

Trace and cut out several of each shape.
Tell whether the shape tessellates. If it does not, explain why.

1. [parallelogram with angles 120°, 60°, 60°, 120°]

_____

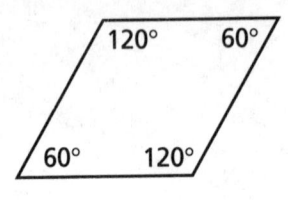

2. [letter Y shape]

_____

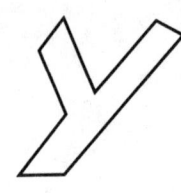

3. [semicircle]

_____

4. [shape like a D with a curved notch]

_____

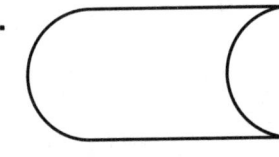

5. [Z-shaped figure]

_____

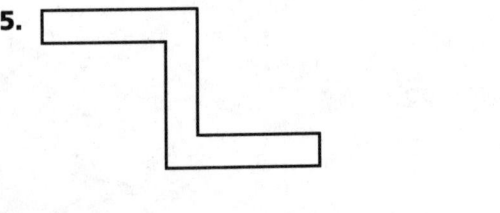

6. [circle with a chord]

_____

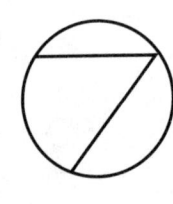

7. Explain why a circle does not tessellate.

_____

8. Will a regular octagon always tessellate? Explain.

_____
_____
_____

9. Will a regular hexagon always tessellate? Explain.

_____
_____
_____

Use with Grade 6, Chapter 20, Lesson 6, pages 500–501.

Name _____

Perimeter

21-1 PRACTICE

Find each perimeter.

1. 7 cm, 4 cm, 3 cm _____

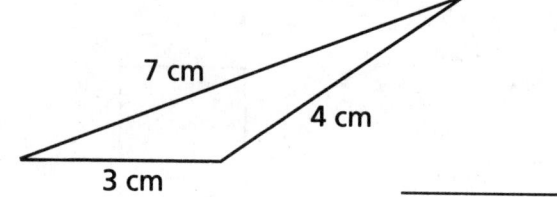

2. 7 m, 5 m _____

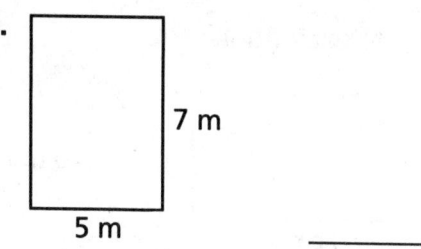

3. 4 in., 8 in., 4 in., 10 in., 6 in., 5 in. _____

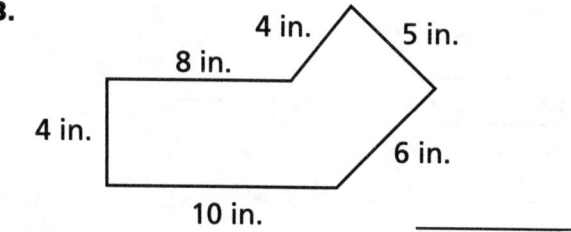

4. 3 cm, 3 cm, 3 cm, 5 cm, 7 cm _____

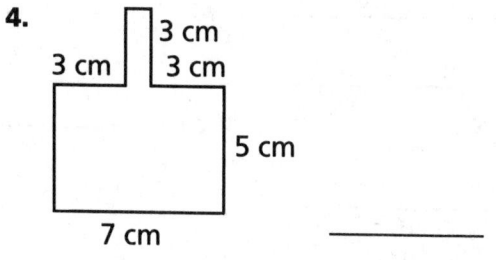

5. 3 cm, 4 cm, 4 cm, 4 cm, 4 cm, 4 cm, 4 cm, 3 cm _____

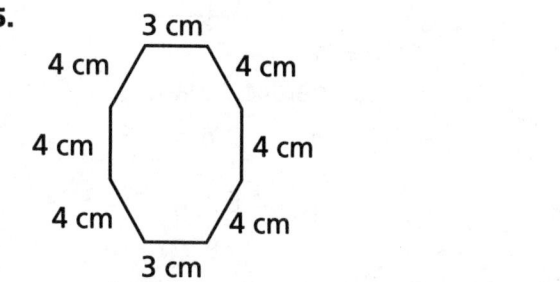

6. 2 in., 2 in., 2 in., 6 in. _____

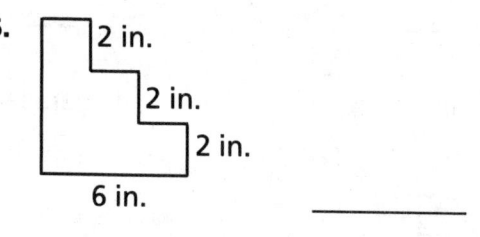

7. 12 ft, 5 ft _____

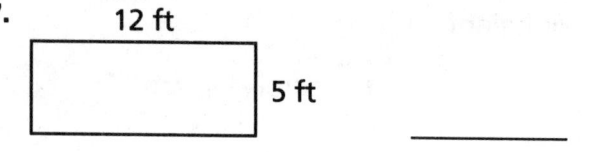

8. 9 m, 9 m _____

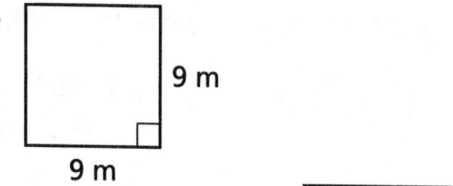

Problem Solving
Solve.

9. Find the perimeter of an isosceles triangle whose sides are 8 inches and whose base is 4 inches.

_____

10. Molly has 60 feet of fencing to go around the perimeter of her garden. She wants the garden to be a square. How long should each side be?

_____

Use with Grade 6, Chapter 21, Lesson 1, pages 516–519.

115

Name _____

Area of Rectangles and Parallelograms • Algebra

P 21-2 PRACTICE

Find the area of each figure.

1. 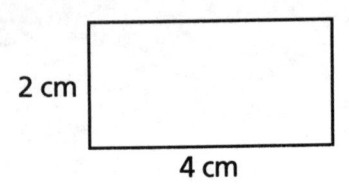 2 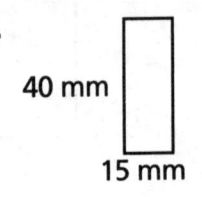 cm, 4 cm _____

2. 40 mm, 15 mm _____

3. 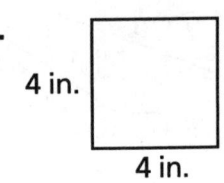 4 in., 4 in. _____

4. 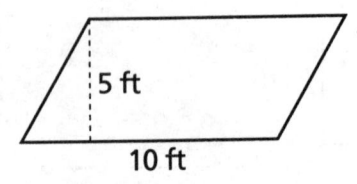 5 ft, 10 ft _____

5. 1.8 cm, 8.4 cm _____

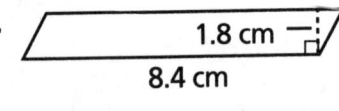

6. 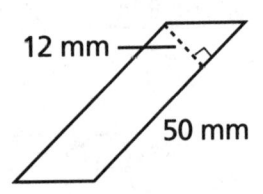 12 mm, 50 mm _____

7. rectangle
   $l = 3\frac{1}{4}$ yd
   $w = 4\frac{3}{4}$ yd
   _____

8. parallelogram
   $b = 4$ in.
   $h = 5$ in.
   _____

9. parallelogram
   $b = 3.2$ mm
   $h = 4.6$ mm
   _____

Algebra Find the unknown length, width, base, or height.

10. rectangle
    $l = 3$ in.
    $A = 6$ in.²
    $w =$ _____

11. rectangle
    $l = 45$ mm
    $A = 3,150$ mm²
    $w =$ _____

12. parallelogram
    $h = 15$ yd
    $A = 675$ yd²
    $b =$ _____

Problem Solving
Solve.

13. Mike's room is $12\frac{1}{2}$ feet by $15\frac{1}{2}$ feet. How many square feet of carpeting does he need to cover the entire floor?

    _____

14. Helen is planting tomatoes in her garden. She can place 3 plants per square foot. How many plants does she need if her garden measures 7 ft by 6 ft?

    _____

Use with Grade 6, Chapter 21, Lesson 2, pages 520–523.

Name _____

Area of Triangles and Trapezoids

P 21-3 PRACTICE

Find the area of each figure.

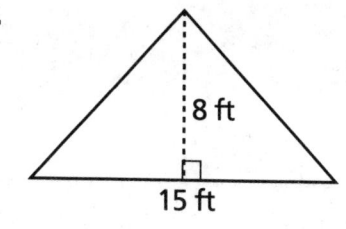

1. Triangle with legs 3 ft and 4 ft, hypotenuse 5 ft.

_____

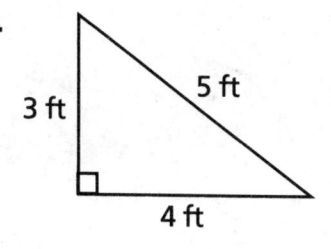

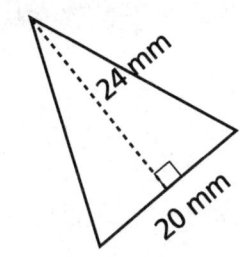

2. Triangle with base 15 ft and height 8 ft.

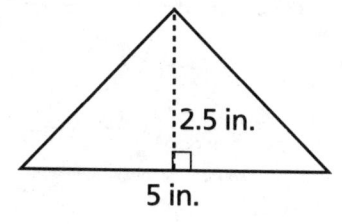

_____

3. Triangle with height 24 mm and base 20 mm.

_____

4. Triangle with base 5 in. and height 2.5 in.

_____

5. Triangle with base 9.2 cm and height 3.4 cm.

_____

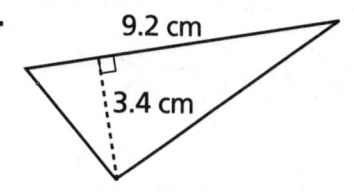

6. Triangle with base $3\frac{3}{4}$ in. and height $2\frac{1}{2}$ in.

_____

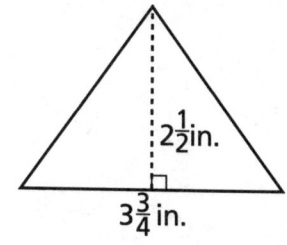

7. Trapezoid with parallel sides 5 cm and 7 cm, height 4 cm.

_____

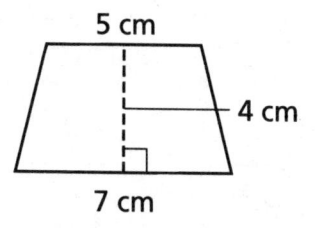

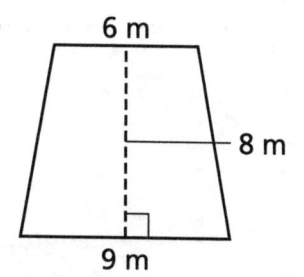

8. Trapezoid with parallel sides 6 m and 9 m, height 8 m.

_____

9. Trapezoid with parallel sides 4 in. and 5 in., height 6 in.

_____

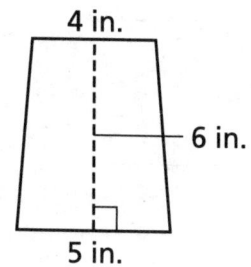

10. Trapezoid with parallel sides 5 ft and 9 ft, height 2 ft.

_____

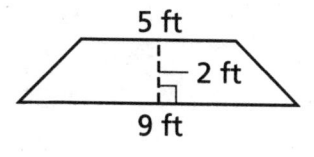

11. Trapezoid with parallel sides 8 yd and 10 yd, height 5 yd.

_____

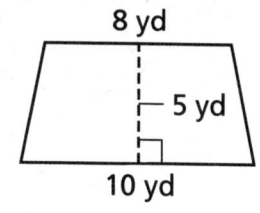

12. Trapezoid with parallel sides 4 m and 6 m, height 3 m.

_____

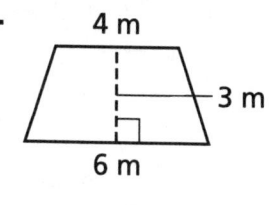

Problem Solving
Solve.

13. The height of a triangle is 28 ft. If the base is 12 ft, what is the area of the triangle?

_____

14. The area of a triangle is 3.99 cm². If the base is 4.2 cm, find its height.

_____

Use with Grade 6, Chapter 21, Lesson 3, pages 524–526.

117

Name_____

Explore Circumference of Circles

P 21-4 PRACTICE

Find the circumference. Use $\pi \approx 3.14$ or $\pi \approx 3\frac{1}{7}$.
Round to the nearest whole number.

1. 4.5 cm _____

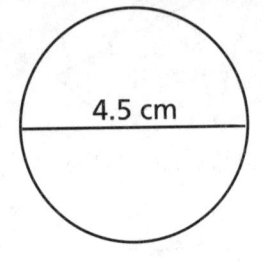

2. 2 in. _____

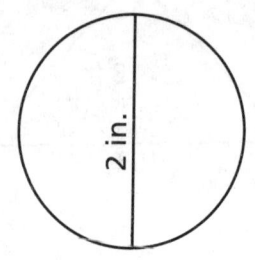

3. $10\frac{1}{2}$ ft _____

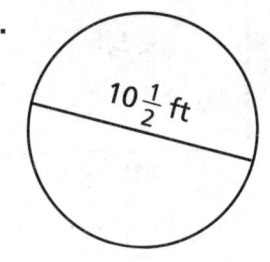

4. 8.5 mm _____

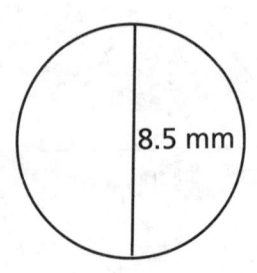

5. 25 in. _____

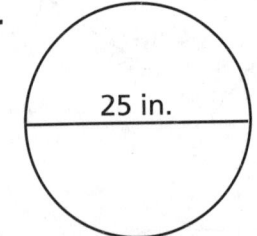

6. $35\frac{1}{2}$ in. _____

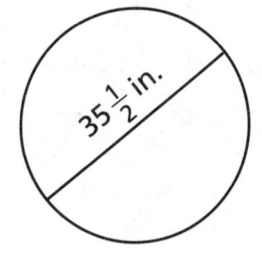

7. diameter = 4 meters _____

8. diameter = 3.7 inches _____

9. diameter = 15 inches _____

10. diameter = 7.8 meters _____

11. diameter = 12.4 centimeters _____

12. diameter = 3.6 feet _____

13. diameter = 32 millimeters _____

14. diameter = 1.25 inches _____

15. diameter = $\frac{3}{4}$ inch _____

Problem Solving
Solve.

16. A town has a circular bicycle path around it. The diameter of the circular path is 262 meters. What is the circumference of the circle? Use $\pi \approx 3.14$. Round to the nearest whole number.

17. A park has a circular pool. The diameter of the pool is 30.5 meters. What is the circumference of the pool? Use $\pi \approx 3.14$. Round to the nearest whole number.

Name _____

Explore Area of Circles

21-5 PRACTICE

Find the area. Use π ≈ 3.14. Round to the nearest hundredth.

1. r = 4 in. _____

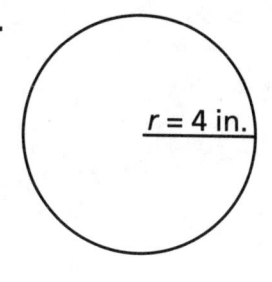

2. r = 7.5 cm _____

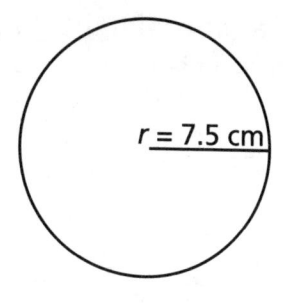

3. r = 0.2 mm _____

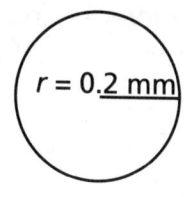

4. r = 25 ft _____

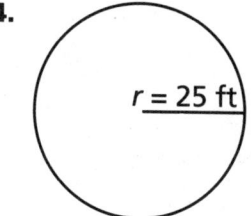

5. r = 1.2 m _____

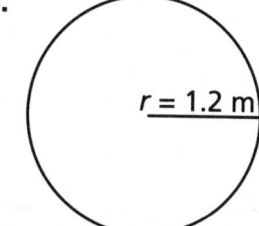

6. r = 4.6 km _____

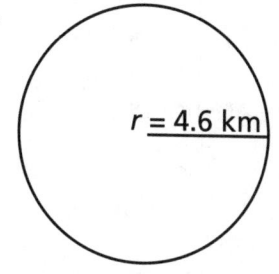

7. radius = 3 inches _____

8. radius = 6 centimeters _____

9. radius = 15 millimeters _____

10. radius = 7 feet _____

11. radius = 0.15 kilometer _____

12. radius = 12 yards _____

13. radius = 4 feet _____

14. radius = 3 meters _____

15. radius = 8 inches _____

16. radius = 5 yards _____

Problem Solving

Solve.

17. A town is planning to make a circular park with a radius of 150 feet. What will the area of the park be? Use π ≈ 3.14.

18. A town is planning to build a circular pool with a radius of about 11 feet. What will the area of the pool be? Use π ≈ 3.14.

Use with Grade 6, Chapter 21, Lesson 5, pages 530–531.

Circumference and Area of Circles • Algebra

21-6 PRACTICE

Find the circumference and the area of a circle with the given radius. Use $\pi \approx 3.14$ or $\pi \approx 3\frac{1}{7}$. Round to the nearest tenth.

1. $r = 18$ meters _____
2. $r = 25$ centimeters _____
3. $r = 7$ decimeters _____
4. $r = 16$ inches _____
5. $r = 12$ yards _____
6. $r = 11$ millimeters _____

Find the area of each circle. Use $\pi \approx 3.14$. Express your answer using significant digits.

7. $r = 1.7$ centimeters _____
8. $d = 21$ inches _____
9. $r = 4.5$ meters _____
10. $d = 11$ centimeters _____
11. $r = 2$ inches _____
12. $d = 12$ meters _____

Algebra Find the radius of a circle with the given circumference. Use $\pi = 3.14$ or $\pi \approx 3\frac{1}{7}$. Round to the nearest hundredth.

13. $C = 113.09$ meters _____   14. $C = 4$ yards _____
15. $C = 43.96$ centimeters _____   16. $C = 12$ inches _____

Problem Solving
Solve.

17. The students in Mr. Smith's science class are planning a circular flower garden with a diameter of 15.4 meters. What will the area of the garden be? Use $\pi \approx 3.14$.

18. The diameter of Earth at the equator is about 12,756 km. What is the circumference of Earth at the equator? Use $\pi \approx 3.14$.

Name _____

Problem Solving: Skill
Solve a Simpler Problem

P 21-7 PRACTICE

Use data from the diagram for problems 1–2. Use $\pi \approx 3.14$.

1. The diagram shows a plan for Meadow Park. The city requires all areas to be covered in grass except for the pond, the playground, and the picnic area. How much grass will be needed to cover the ground in the park?

   _____

2. The plans call for the picnic area to have a wooden floor. How much wood is needed for the picnic area?

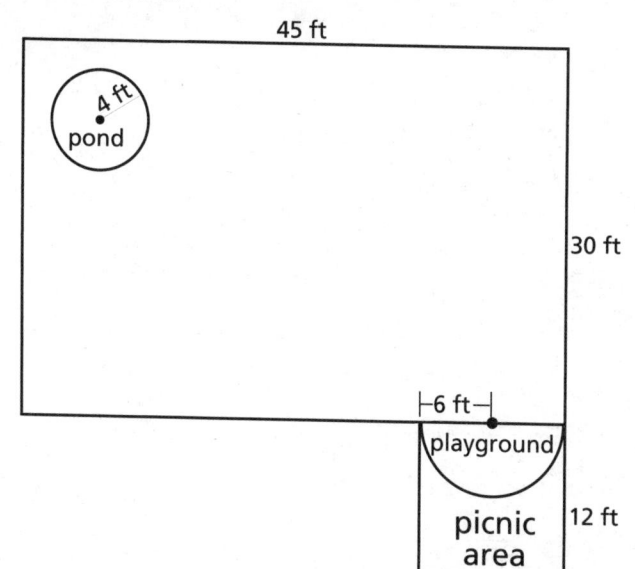

   _____

(Diagram: rectangle 45 ft by 30 ft with a pond of radius 4 ft inside; a semicircular playground of diameter 6 ft on the bottom edge; a picnic area 12 ft below playground)

Choose the correct answer.

3. A banner that is 15 inches long and 9 inches wide has a circular cutout with a radius of 3 inches. What is the area of the banner?

   A 107 in.²   C 126 in²
   B 135 in²   D 163 in.²

4. A circular field has a radius of 6 meters. A shed that is 2 meters wide and 4 meters long stands inside the field. What is the area of the field that is not covered by the shed?

   F 28 m²   H 105 m²
   G 113 m²   I 121 m²

Mixed Strategy Review
Solve. Use any strategy.

5. The girls' basketball team is having a bake sale. They are selling 2 brownies for $1.50. How much money will they make for the team if they sell 96 brownies?

   _____

   Strategy: _____

6. Denise wants to go to a movie that begins at 7:30 P.M. It takes her 15 minutes to get to the theater. She must finish her homework before she can go. Her homework will take her 40 minutes to complete. What is the latest time that she should start her homework to arrive on time for the movie?

   _____

   Strategy: _____

Use with Grade 6, Chapter 21, Lesson 7, pages 536–537.

Name _____

3-Dimensional Figures

P 22-1 PRACTICE

Classify each figure. If it is a polyhedron, find the number of faces, edges, and vertices.

1.

_____
_____

2.

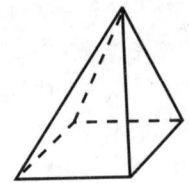

_____
_____

3.

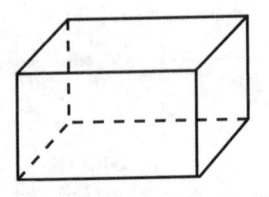

_____
_____

4.

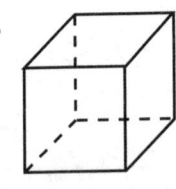

_____
_____

5.

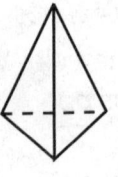

_____
_____

6.

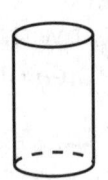

_____
_____

Draw each figure.

7. cylinder

8. rectangular prism

9. triangular prism

10. cone

11. square pyramid

12. cube

Use with Grade 6, Chapter 22, Lesson 1, pages 542–545.

Name_____

Surface Area of Prisms

P 22-2 PRACTICE

Find the surface area of each figure.

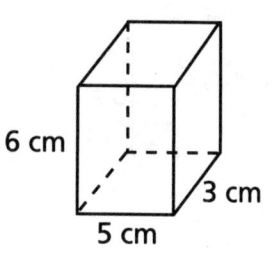

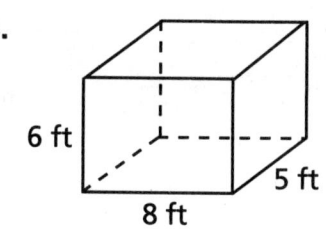

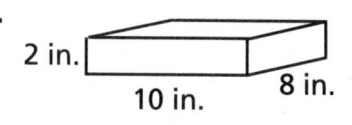

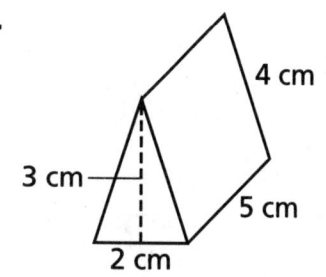

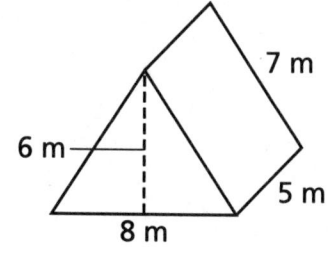

1. 4 in. × 4 in. × 4 in. cube

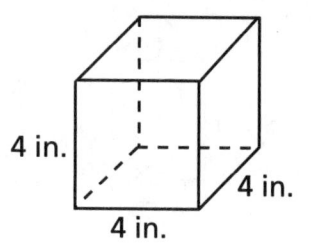

_____

2. 6 cm × 5 cm × 3 cm rectangular prism

_____

3. 6 ft × 8 ft × 5 ft rectangular prism

_____

4. 2 in. × 10 in. × 8 in. rectangular prism

_____

5. Triangular prism: 4 cm, 3 cm, 2 cm, 5 cm

_____

6. Triangular prism: 7 m, 6 m, 8 m, 5 m

_____

7. 4 yd × 5 yd × 4 yd rectangular prism

_____

8. Triangular prism: 8 m, 4 m, 6 m, 10 m

_____

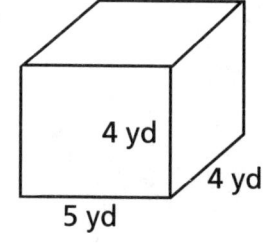

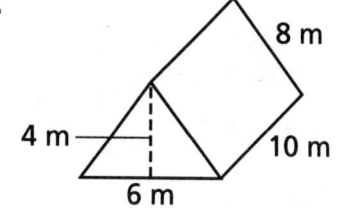

9. 12 ft × 12 ft × 12 ft cube

_____

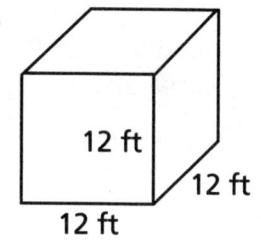

Problem Solving
Solve.

10. Corinne is painting a box that will be used as a prop in a play. The box is 6 feet by 5 feet by 3 feet. Find the surface area of the box.

_____

11. Andrew is preparing another box that will be used as a prop. The box is $8\frac{1}{2}$ feet by $6\frac{1}{2}$ feet by $4\frac{1}{2}$ feet. Find the surface area of this box.

_____

Use with Grade 6, Chapter 22, Lesson 2, pages 546–548.

123

Name _____

Volume of Prisms • Algebra

22-3 PRACTICE

Find the volume of each figure.

1. 2 in., 10 in., 8 in. _____

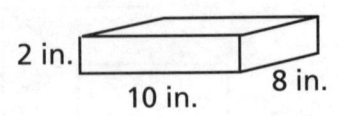

2. 6 ft, 4 ft, 3 ft _____

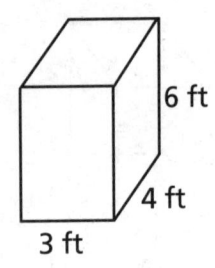

3. 9 cm, 3 cm, 6 cm _____

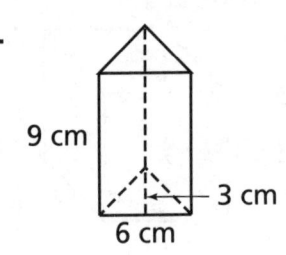

4. 14 yd, 6 yd, 10 yd _____

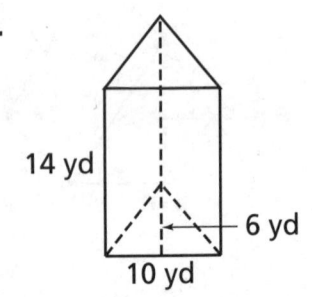

5. 7 m, 5 m, 8 m _____

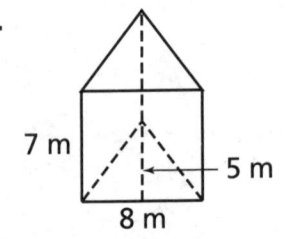

6. 12 ft, 12 ft, 12 ft _____

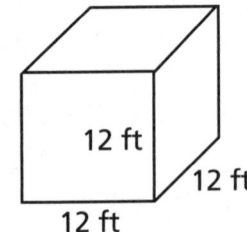

Find the volume of each rectangular prism.
Round decimal answers to the nearest tenth.

7. $l = 12$ ft
   $w = 9$ ft
   $h = 6$ ft

8. $l = 4.2$ m
   $w = 3.8$ m
   $h = 2.6$ m

9. $l = 3.3$ cm
   $w = 1.4$ cm
   $h = 3.5$ cm

Problem Solving
Solve.

10. Shannon needs to fill a box so it won't tip over. The box is 6 feet by 5 feet by 3 feet. Find the volume of the box.

11. Wayne wants to fill another box that is $8\frac{1}{2}$ feet by $6\frac{1}{2}$ feet by $4\frac{1}{2}$ feet. Find the volume of this box.

124

Use with Grade 6, Chapter 22, Lesson 3, pages 550–553.

Name _____

Problem Solving: Strategy
Logical Reasoning

P 22-4 PRACTICE

A pool is 80 feet long, 25 feet wide, and 12 feet deep. For each problem, decide if you need to find the perimeter, area, or volume. Then solve.

1. The bottom of the pool is sealed with a plastic coating. What size coating would be needed to cover the entire bottom of the pool?

   _____

2. The pool is half full. How much more water is needed to fill the pool?

   _____

3. A tile border lines the outside of the pool. How long is the tile border?

   _____

4. What size cover is needed to cover the pool when it is empty?

   _____

Mixed Strategy Review
Solve. Use any strategy.

5. Kim makes four prints using a pattern. The largest print has an area of 120 in.² The next two prints have areas of 60 in.² and 30 in.² What is the area of the smallest print?

   _____

   Strategy: _____

6. Jeff is designing the photo page of the yearbook. The pages are 12 inches by 9 inches. Each photo needs 2 inches by 3 inches of space on the page. How many photos will fit on one page?

   _____

   Strategy: _____

7. What two positive integers have a sum of 12 and a product of 32?

   _____

   Strategy: _____

8. Orlando has 3 more than twice as many shells in his collection as Kimberly has. How many shells does Orlando have if Kimberly has 28 shells in her collection?

   _____

   Strategy: _____

Use with Grade 6, Chapter 22, Lesson 4, pages 554–555.

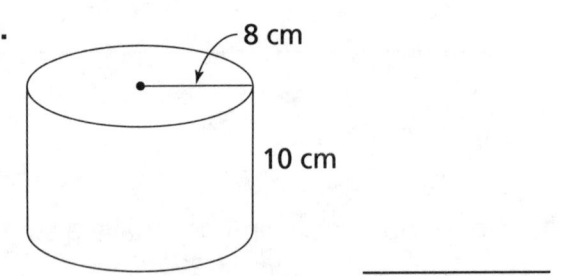

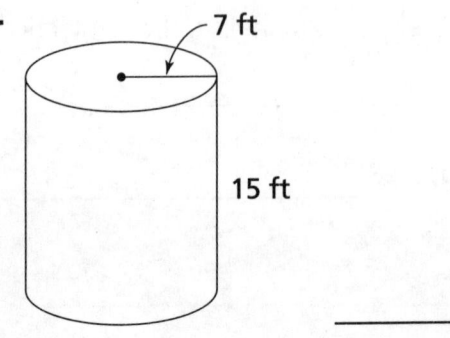

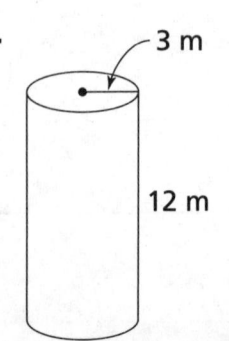

Name_____

Explore Surface Area of Cylinders • Algebra   22-5 PRACTICE

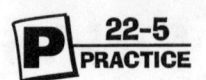

Find the surface area. Use π ≈ 3.14.

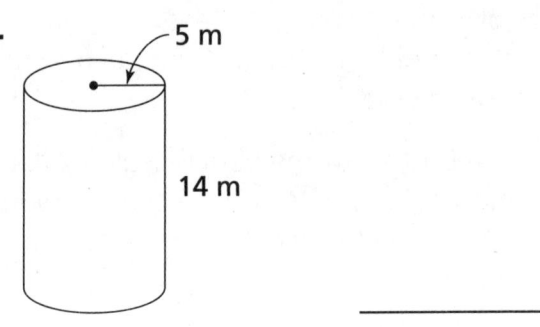

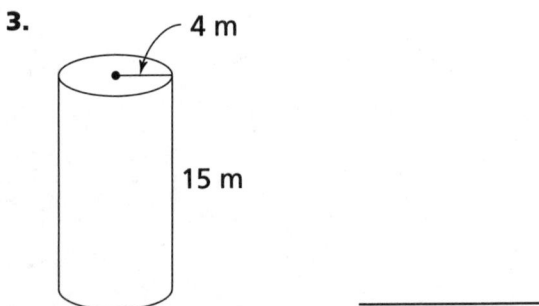

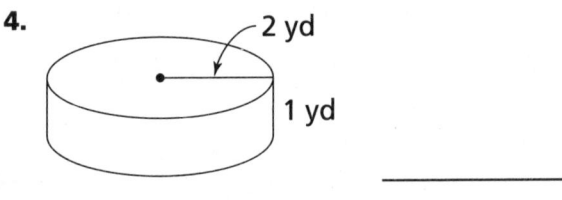

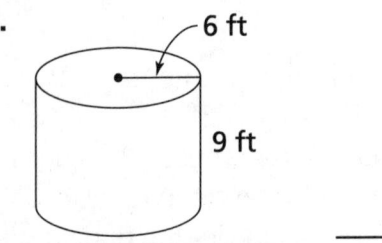

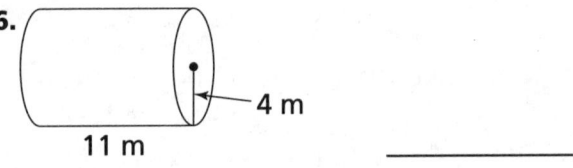

1. 3 in. (radius), 2 in. (height) _____

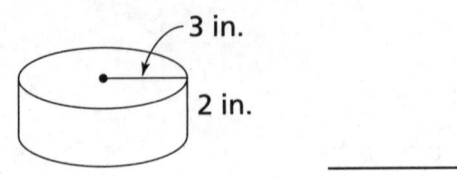

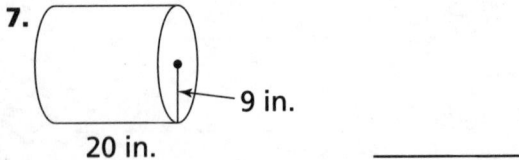

2. 5 m (radius), 14 m (height) _____

3. 4 m (radius), 15 m (height) _____

4. 2 yd (radius), 1 yd (height) _____

5. 6 ft (radius), 9 ft (height) _____

6. 4 m (radius), 11 m (height) _____

7. 9 in. (radius), 20 in. (height) _____

8. 8 cm (radius), 10 cm (height) _____

9. 7 ft (radius), 15 ft (height) _____

10. 3 m (radius), 12 m (height) _____

126   Use with Grade 6, Chapter 22, Lesson 5, pages 556–557.

Name _____

Explore Volume of Cylinders

P 22-6 PRACTICE

Find the volume. Use π ≈ 3.14.

1. 5 ft, 15 ft

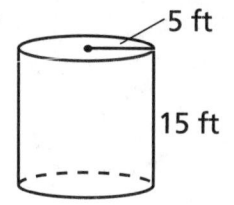

2. 32 cm, 150 cm

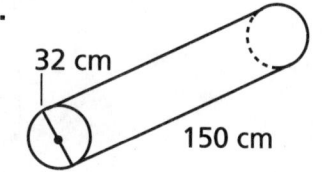

3. 6 mm, 33 mm

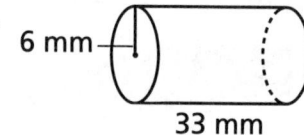

4. 4 yd, 3 yd

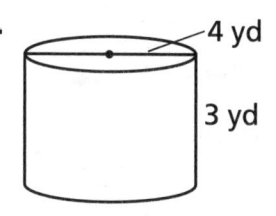

5. 3.5 in., 4.8 in.

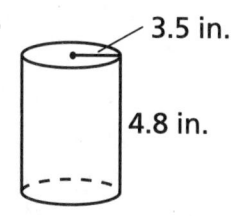

6. 1.8 m, 15.5 m

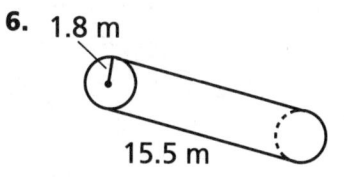

7. 9 m, 6 m

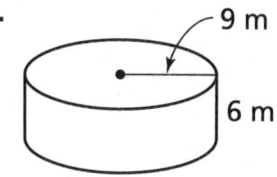

8. 4 ft, 15 ft

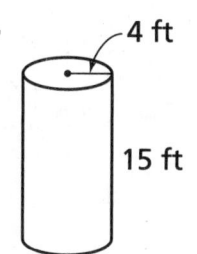

9. 3 in., 10 in.

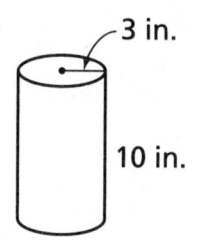

Find the volume. Round to the nearest thousandth. Use π ≈ 3.14.

10. diameter: 4.2 centimeters
    height: 8.2 centimeters

11. diameter: 4 centimeters
    height: 4 centimeters

12. diameter: 6 meters
    height: 4.3 meters

13. diameter: 2 centimeters
    height: 3.8 centimeters

14. diameter: 6.2 meters
    height: 9 meters

15. diameter: 8.4 meters
    height: 1.3 meters

Use with Grade 6, Chapter 22, Lesson 6, pages 558–559.

Name_____

Ratios and Equivalent Ratios

P 23-1 PRACTICE

Write a ratio comparing the shaded region to the unshaded region.
Then write a ratio comparing the shaded region to the whole figure.

1.

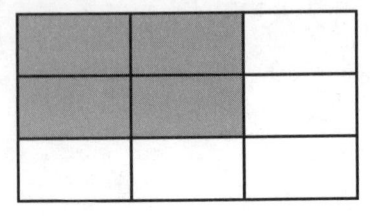

_____

2.

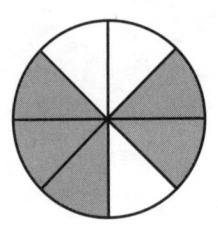

_____

Compare. Write >, <, or =.

3. $\frac{3}{1}$ and $\frac{9}{3}$ _____

4. $\frac{3}{5}$ and $\frac{7}{12}$ _____

5. $\frac{2}{5}$ and $\frac{6}{15}$ _____

6. $\frac{2}{5}$ and $\frac{8}{25}$ _____

7. $\frac{18}{63}$ and $\frac{2}{7}$ _____

8. $\frac{15}{25}$ and $\frac{3}{5}$ _____

9. $\frac{24}{48}$ and $\frac{1}{2}$ _____

10. $\frac{1}{3}$ and $\frac{3}{9}$ _____

11. $\frac{12}{32}$ and $\frac{3}{8}$ _____

12. $\frac{4}{5}$ and $\frac{16}{25}$ _____

13. $\frac{1}{4}$ and $\frac{7}{28}$ _____

14. $\frac{5}{9}$ and $\frac{15}{18}$ _____

Algebra Find each unknown number.

15. $\frac{2}{3} = \frac{n}{6}$ _____

16. $\frac{5}{6} = \frac{x}{36}$ _____

17. $\frac{3}{8} = \frac{y}{24}$ _____

18. $\frac{5}{7} = \frac{a}{42}$ _____

19. $\frac{3}{8} = \frac{a}{40}$ _____

20. $\frac{4}{5} = \frac{36}{n}$ _____

21. $\frac{r}{6} = \frac{12}{18}$ _____

22. $\frac{b}{2} = \frac{7}{14}$ _____

23. $\frac{3}{5} = \frac{m}{30}$ _____

24. $\frac{7}{10} = \frac{21}{c}$ _____

Problem Solving
Solve.

25. The ratio of green stripes to yellow stripes on a flag is 4 to 7. Write this ratio in two other ways.

_____

26. The ratio of the width to the length of a flag is 9 to 10. The ratio of the width to the length of another flag is 18 to 20. Are these two ratios equivalent? Explain.

_____

128

Use with Grade 6, Chapter 23, Lesson 1, pages 574–576.

Name _____

Rates

P 23-2 PRACTICE

Find each unit rate.

1. Potatoes: $1.00 for 4 kilograms = _____ for 1 kilogram

2. Carol's truck: 250 miles on 10 gallons = _____ on 1 gallon

3. Doughnuts: $1.56 for 1 dozen = _____ for 1 doughnut

4. 24 pictures for $12.00 = _____ for 1 picture

5. 48 baseballs in 6 boxes = _____ baseballs in 1 box

6. 600 people per 15 square miles = _____ people per square mile

7. 360 miles in 4 hours = _____ miles in 1 hour

8. $200 for 5 days = _____ for 1 day

9. 24 feet in 6 hours = _____ feet in 1 hour

10. 64 meters in 8 seconds = _____ meters in 1 second

11. 75 pages in 3 hours = _____ pages per 1 hour

12. $496 in 4 days = _____ in 1 day

Algebra Kyle and Natalie drove 208 miles in 4 hours. Use their rate to solve problems 13–16.

13. How many miles can they drive in 1 hour? _____

14. How many miles can they drive in 6 hours? _____

15. How many hours will it take them to drive 416 miles? _____

16. How many hours will it take them to drive 624 miles? _____

Problem Solving
Solve.

17. A car uses 15 gallons of gas to travel 405 miles. How many miles per gallon does the car get?

18. Mel hiked 54 miles in 3 days. At the same rate, how many miles can Mel hike in 5 days?

_____   _____

Use with Grade 6, Chapter 23, Lesson 2, pages 578–579.

129

Name _____

Better Buy

P 23-3 PRACTICE

Find each unit price. Round to the nearest cent.

1. 4 cookies for 99¢ _____
2. 10 cookies for $2.75 _____
2. 12 note pads for $15.97 _____
4. 5 note pads for $6.75 _____
5. 15 yards of ribbon for $23.99 _____
6. 40 yards of ribbon for $59.90 _____
7. 25 pens for $16.50 _____
8. 80 pens for $51.50 _____

Find the better buy.

9. fruit drink: 64 ounces for $2.39 or 20 ounces for $1.05

10. frozen cheese pizza: 21 ounces for $4.59 or 10 ounces for $1.59

_____

_____

11. peanut butter: 18 ounces for $1.99 or 28 ounces for $3.59

12. orange juice: 1 gallon for $4.50 or 1 quart for $1.25

_____

_____

13. grapefruit: 6 for $1.38 or 8 for $1.75

14. onions: 3 pounds for $0.95 or 10 pounds for $3.25

_____

_____

15. granola: 16 ounces for $2.88 or 24 ounces for $4.40

16. turkey: 2 pounds for $4.98 or 3 pounds for $7.55

_____

_____

Algebra Find *x* in each problem.

17. A bag of 9 oranges costing $2.88 has a unit price of *x* per orange.

18. A 32-ounce jar of tomato sauce that costs *x* has a unit price of $0.12 per ounce.

_____

_____

Problem Solving
Solve.

19. Harry needs 12 ounces of chocolate chips for a dessert he is making. A 10-ounce bag of chips costs $2.00, and a 24-ounce bag costs $4.19. What should he buy? Explain.

20. At one store, the price of film is 3 rolls for $15.39. At another, the same film is 5 rolls for $24.99. Which is the better buy?

_____

_____

130

Use with Grade 6, Chapter 23, Lesson 3, pages 580–581.

Name_____

Problem Solving: Skill
Check for Reasonableness

P 23-4 PRACTICE

Use data from the table for problems 1–2. Check whether each answer is reasonable. Then solve.

Lori competes in bicycle road races. The table shows a record of Lori's race times.

Race	1	2	3	4
Distance	50 miles	60 miles	100 miles	80 miles
Time	3.8 hours	4.9 hours	11.8 hours	7.4 hours

1. Lori calculates that her unit rate for the second race was about 18 miles per hour. Is her calculation reasonable? Explain.

   _____
   _____

2. Lori's brother calculates that her unit rate for the fourth race was about 11 miles per hour. Is his calculation reasonable? Explain.

   _____
   _____

Choose the correct answer.
Erin is traveling with her family. She has been in the car for 4 hours. Her mother says they have traveled 192 miles. Erin calculates that they are driving at an average rate of 36 miles per hour.

3. Which of the following statements is true?

   A Erin's family is traveling at exactly 36 miles per hour.

   B Erin's calculation is reasonable.

   C A better calculation would be about 50 miles per hour.

   D Erin's family drove 192 miles in 36 hours.

4. To check if an answer is reasonable, you

   F compare facts with a guess.

   G perform at least two operations.

   H always estimate the answer.

   J compare the answer with what you know.

Mixed Strategy Review

5. A rancher is building a square corral with sides that are 20 feet long. He plans to put a post every 5 feet around the edge of the corral. How many posts will he need?

   _____

6. At 5 P.M., the temperature was 3°C. By 8 P.M., the temperature had dropped 6°C. What was the temperature at 8 P.M.?

   _____

Use with Grade 6, Chapter 23, Lesson 4, pages 582–583.

131

Name_____

Proportions • Algebra

24-1 PRACTICE

Solve each proportion.

1. $\dfrac{n}{6} = \dfrac{6}{9}$ _____

2. $\dfrac{10}{n} = \dfrac{4}{8}$ _____

3. $\dfrac{4}{8} = \dfrac{2}{n}$ _____

4. $\dfrac{n}{6} = \dfrac{15}{45}$ _____

5. $\dfrac{2}{10} = \dfrac{n}{35}$ _____

6. $\dfrac{15}{7} = \dfrac{n}{105}$ _____

7. $\dfrac{12}{13} = \dfrac{n}{130}$ _____

8. $\dfrac{21}{6} = \dfrac{35}{n}$ _____

9. $\dfrac{n}{8} = \dfrac{3}{4}$ _____

10. $\dfrac{n}{10} = \dfrac{40}{1.6}$ _____

11. $\dfrac{4.9}{n} = \dfrac{28}{8}$ _____

12. $\dfrac{9}{100} = \dfrac{n}{50}$ _____

13. $\dfrac{9}{n} = \dfrac{54}{12}$ _____

14. $\dfrac{5}{3.4} = \dfrac{2.5}{n}$ _____

15. $\dfrac{n}{2.4} = \dfrac{26}{52}$ _____

16. $\dfrac{n}{8} = \dfrac{2.25}{6}$ _____

17. $\dfrac{2.5}{5} = \dfrac{n}{10}$ _____

18. $\dfrac{n}{2} = \dfrac{11}{5}$ _____

Write a proportion and solve.

19. Jeanette can walk 1 km in 11 minutes. At the same rate, how far can she walk in 55 minutes?

20. On Mickey's block the ratio of trees to houses is 19:5. If there are 15 houses, how many trees are there?

21. Carlos can ride his bike 30 km in 2 hours. At the same rate, how far can he ride in 7 hours?

22. At Sarah's school the ratio of students to teachers is 94 to 3. If there are 752 students in the school, how many teachers are there?

23. On George's block the ratio of cars to houses is 8 to 5. There are 32 cars. How many houses are there?

24. A machine can pack sneakers at a rate of 4 pairs every 12 seconds. How many pairs of sneakers can this machine pack in 1 minute?

Name _____

Problem Solving: Strategy
Make a Graph

P 24-2 PRACTICE

Write an equation. Solve.

1. A recipe for punch uses juice and sparkling water in a ratio of 2 to 5. If Gail mixes in 6 pints of juice, how many pints of sparkling water should she use?

   _____

2. The ratio of wax to oil paint that an artist uses to thicken paint is 3 to 10. If the artist has 60 ounces of paint, how many ounces of wax does he need?

   _____

3. A baker makes a mixed grain bread using wheat flour and white flour in a ratio of 3 to 4. How many cups of wheat flour does the baker need for 20 cups of white flour?

   _____

4. The ratio of girls to boys on a field trip is 8 to 7. If there are 56 girls on the trip, how many students in all are on the field trip?

   _____

Mixed Strategy Review
Solve. Use any strategy.

5. Victor is building a fence around a rectangular play area. The area measures 10 yards by 6 yards. He must dig postholes at the corners and every 2 yards along the perimeter. How many postholes does Victor have to dig?

   _____

   Strategy: _____

6. Melissa surveyed 15 of her friends and found that 10 have brothers and 8 have sisters. How many of Melissa's friends have both sisters and brothers?

   _____

   Strategy: _____

7. Frances uses a 2-to-5 ratio of orange juice to strawberry juice to make fruit punch. If she wants to make 35 cups of fruit punch, how many cups of orange juice and strawberry juice does she need?

   _____
   _____

   Strategy: _____

8. **Write a problem** that you could use a graph to solve. Share it with others.

   _____
   _____
   _____

Use with Grade 6, Chapter 24, Lesson 2, pages 592–593.

133

Name _____

Explore Similar Figures

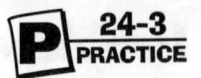

24-3 PRACTICE

triangle *SRP* ~ triangle *NLM*

Find each corresponding side.

1. side *SP* _____
2. side *RP* _____
3. side *RS* _____

Find each corresponding angle.

4. ∠*P* _____
5. ∠*R* _____
6. ∠*S* _____

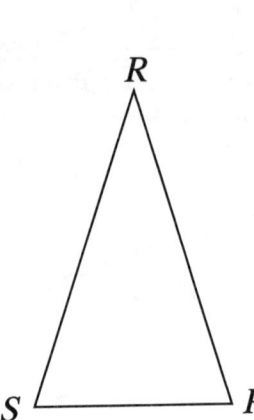

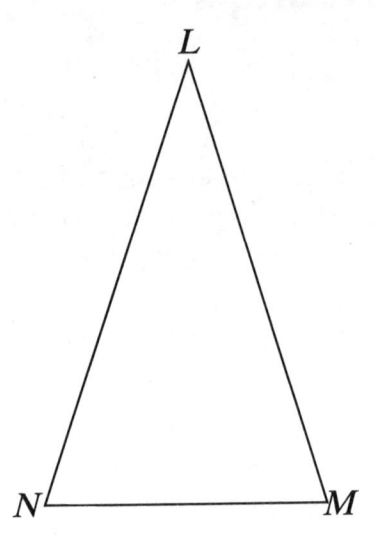

Find the unknown length for each pair of similar figures.

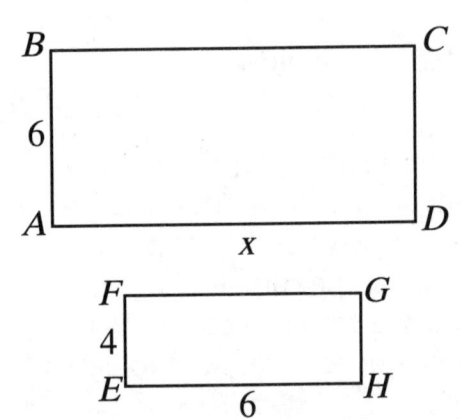

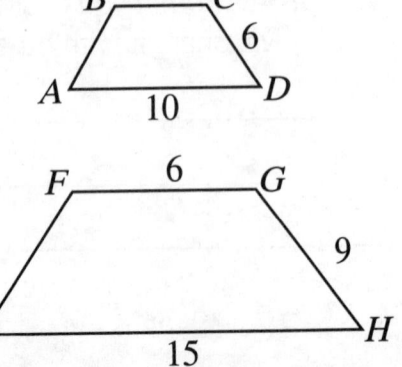

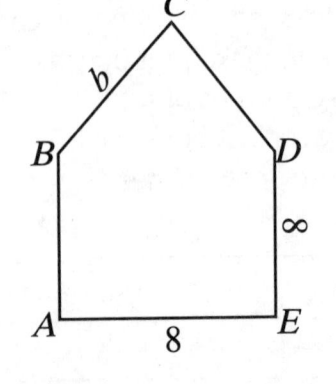

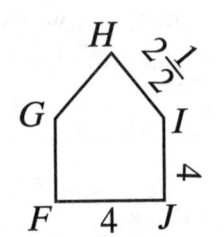

7. Triangle ABC with y, 5, $3\frac{1}{2}$; Triangle DEF with 10, 10, 7.

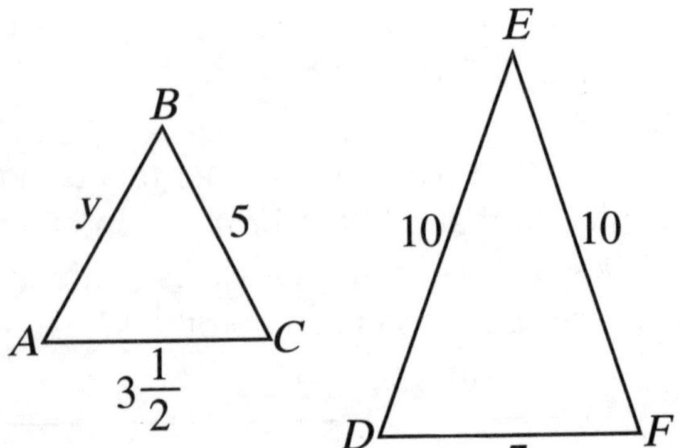

8. Rectangle ABCD with 6, x; Rectangle EFGH with 4, 6.

9. Trapezoid ABCD with z, 6, 10; Trapezoid EFGH with 6, 9, 15.

10. Pentagon with b, 8, 8; Pentagon with $2\frac{1}{2}$, 4, 4.

134

Use with Grade 6, Chapter 24, Lesson 3, pages 594–595.

Name _____

Scale Drawings and Maps

P 24-4 PRACTICE

Use the scale 1 inch = 6 feet to find the actual distance for each scale distance.

1. 3 inches _____
2. 5 inches _____
3. 12 inches _____
4. 20 inches _____

5. 38 inches _____
6. 64 inches _____
7. $2\frac{1}{2}$ inches _____
8. $3\frac{1}{2}$ inches _____

Use the scale 1 inch = 6 feet to find the scale distance for each actual distance.

9. 30 feet _____
10. 48 feet _____
11. 66 feet _____
12. 135 feet _____

13. 18 feet _____
14. 75 feet _____
15. 36 feet _____
16. 228 feet _____

Use the scale 6 centimeters = 30 kilometers to find the actual distance for each scale distance.

17. 10 centimeters _____
18. 25 centimeters _____
19. 30 centimeters _____

20. 38 centimeters _____
21. 92 centimeters _____
22. 14 centimeters _____

Use the scale 6 centimeters = 30 kilometers to find the scale distance for each actual distance.

23. 72 kilometers _____
24. 288 kilometers _____
25. 432 kilometers _____

26. 18 kilometers _____
27. 96 kilometers _____
28. 369 kilometers _____

Problem Solving
Solve.

29. A model for a house uses a scale of 1 cm = 5 m. If the walls of a room in the model measure 2.5 cm long by 4 cm wide, what will be the actual measurements of the wall?

    _____

30. A model for a house uses a scale of 1 cm = 2.5 m. If the height of the actual room measures 3.75 m, how many centimeters high is the room in the model?

    _____

Use with Grade 6, Chapter 24, Lesson 4, pages 596–598.

135

Name _____

Explore the Meaning of Percent

P 25-1 PRACTICE

Use the grids to show each fraction, ratio, or decimal. Tell what percent is shaded. Record the percent.

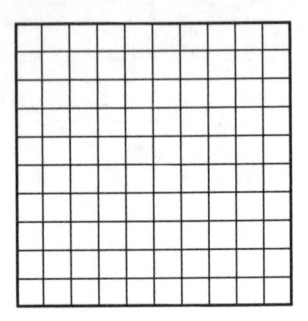

1. 0.42

2. $\frac{1}{2}$

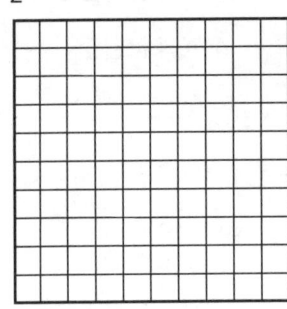

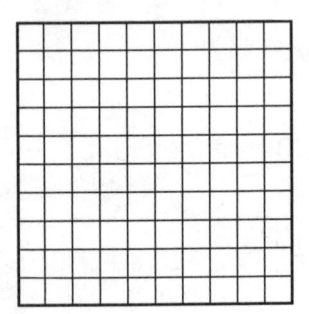

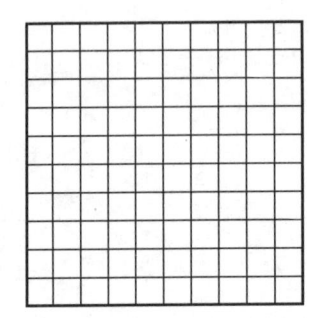

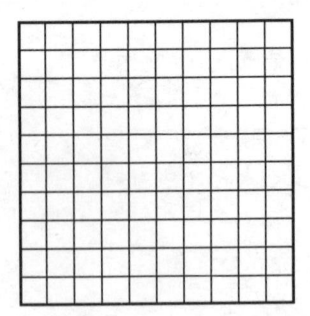

3. 21:100

4. 0.66

5. 27:100

6. $\frac{3}{4}$

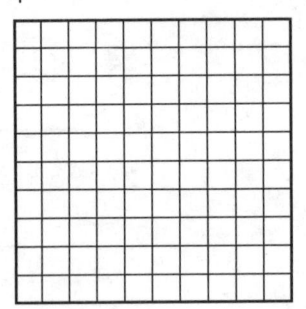

Write each fraction, ratio, or decimal as a percent. You may use 10 by 10 grids to help you.

7. $\frac{2}{5}$ _____

8. 4:20 _____

9. 0.45 _____

10. 0.38 _____

11. $\frac{1}{2}$ _____

12. $\frac{1}{4}$ _____

136

Use with Grade 6, Chapter 25, Lesson 1, pages 614–615.

Name _____

Percents, Fractions, and Decimals

P 25-2 PRACTICE

Write each decimal as a percent.

1. 0.28 = _____
2. 0.13 = _____
3. 0.8 = _____
4. 0.92 = _____
5. 0.43 = _____
6. 0.01 = _____

Write each percent as a decimal.

7. 60% = _____
8. 7% = _____
9. 90% = _____
10. 11% = _____
11. 73% = _____
12. 54% = _____

Write each fraction as a percent.

13. $\frac{65}{100}$ = _____
14. $\frac{19}{50}$ = _____
15. $\frac{3}{25}$ = _____
16. $\frac{3}{10}$ = _____
17. $\frac{1}{20}$ = _____
18. $\frac{11}{20}$ = _____

Write each percent as a fraction in simplest form.

19. 24% = _____
20. 60% = _____
21. 15% = _____
22. 45% = _____
23. 80% = _____
24. 4% = _____

Compare. Write >, <, or =.

25. 15% ◯ 51%
26. 11.6% ◯ 10.9%
27. 96% ◯ 89%
28. 2.5% ◯ 2%
29. 28.2% ◯ 30%
30. 75% ◯ 7.5%

Problem Solving
Solve.

31. Of 250 students surveyed, 128 said that they go to a library at least once a week. What percent of 250 is 128? Write your answer to the nearest tenth of a percent.

32. A music store finds that 15 out of 20 people who buy CDs at the store are under 20 years of age. Write 15 out of 20 as a percent.

Use with Grade 6, Chapter 25, Lesson 2, pages 616–619.

Name _____

Percent of a Number

P 25-3 PRACTICE

Find the percent of each number. Round to the nearest hundredth or cent if necessary.

1. 38% of 64 = _____
2. 2% of 18 = _____
3. 130% of 25 = _____

4. 20% of 35 = _____
5. 50% of $160 = _____
6. 2% of 50 = _____

7. 25% of 60 = _____
8. 27% of $90 = _____
9. 42% of 700 = _____

10. 46% of $72 = _____
11. 20% of $82 = _____
12. 125% of 34 = _____

13. 75% of 59 = _____
14. 150% of 95 = _____
15. 4% of $16.30 = _____

16. 18% of $72 = _____
17. 15% of 180 = _____
18. 75% of 640 = _____

Estimate each tip.

19. 20% of $40 = _____
20. 15% of $68 = _____

21. 15% of $25.95 = _____
22. 20% of $120 = _____

Compare. Write <, >, or =.

23. 12% of 74 ◯ 12% of 54
24. 40% of 68 ◯ 42% of 68

25. 42% of 50 ◯ 50% of 42
26. 25% of 80 ◯ 50% of 40

Problem Solving
Solve.

27. Rita orders lemon chicken for $12.95. The tax is 4.5%. What is her bill, including tax? Estimate a 15% tip.

28. David orders chicken cashew nut for $9.50. The tax is 8%. What is his bill, including tax? Estimate a 20% tip.

29. Linda orders shrimp supreme for $13.75. The tax is 6%. What is her bill, including tax? Estimate a 20% tip.

30. Greg orders rolled beef for $8.95. The tax is 7%. What is his bill, including tax? Estimate a 15% tip.

Name _____

Problem Solving: Skill
Choose a Representation

P 25-4 PRACTICE

Use data from the table for problems 1 and 2.

Temperature Range	60°F–69°F	70°F–79°F	80°F–89°F	90°F–99°F
Number of Days in June	1	15	11	3

1. Mr. Watkins predicted that 50% of the days in June would have temperatures 80°F or above. Did 50% of the days have temperatures 80°F or above? Explain.

2. Mr. Watkins predicted that 95% of the days in June would have temperatures 70°F or above. Did 95% of the days have temperatures 70°F or above? Explain.

Choose the correct answer.

Mr. Johnson predicted that 60% of the days in December would have temperatures 30°F or above. He recorded 21 days in December with temperatures 30°F or above.

3. Which of the following statements is true?

   A Mr. Johnson predicted that $\frac{1}{2}$ the days in December would have temperatures 30°F or above.

   B 60% of the days in December had temperatures 30°F or above.

   C Less than 60% of the days in December had temperatures 30°F or above.

   D More than $\frac{3}{5}$ of the days in December had temperatures 30°F or above.

4. When deciding how to represent a percent, you should

   F always draw a picture.

   G always work with a decimal.

   H work with both fractions and decimals.

   J think about how it will be used.

Mixed Strategy Review
Solve.

5. Debra makes a necklace using a 1 to 8 ratio of red beads to blue beads. If she wants to use 36 beads to make the necklace, how many beads of each color will she need?

6. Bryan wants to cover his bookshelf with shelf paper. The shelf is 8 inches wide and 36 inches long. How much paper does he need to buy?

Use with Grade 6, Chapter 25, Lesson 4, pages 624–625.

139

Name_____

Find the Percent One Number Is of Another   25-5 PRACTICE

Find each percent. Round to the nearest tenth of a percent if necessary.

1. 69 is what percent of 99? _____
2. What percent of 8 is 6? _____
3. What percent of 15 is 12? _____
4. 13 is what percent of 52? _____
5. 10 is what percent of 13? _____
6. What percent of 48 is 18? _____
7. What percent of 11 is 4? _____
8. 35 is what percent of 49? _____
9. What percent of 250 is 75? _____
10. What percent of 4 is 1? _____
11. 72 is what percent of 90? _____
12. 11 is what percent of 19? _____
13. What percent of 8 is 5? _____
14. 30 is what percent of 30? _____
15. What percent of 125 is 25? _____
16. What percent of 90 is 18? _____
17. 7 is what percent of 8? _____
18. 49 is what percent of 50? _____
19. What percent of 9 is 6? _____
20. 6 is what percent of 10? _____
21. 10 is what percent of 14? _____
22. 3 is what percent of 20? _____
23. What percent of 40 is 12? _____
24. What percent of 60 is 48? _____

Problem Solving
Solve.

25. There are 56 members of a hiking club. Of the members, 22 are senior citizens. What percent of the members are senior citizens?

26. A survey of 80 senior citizens found that 48 senior citizens walked more than $\frac{1}{2}$ hour per day. What percent of the senior citizens walked more than $\frac{1}{2}$ hour per day?

Name _____

Sales Tax and Discounts

26-1 Practice

Find each total cost, including sales tax.

1. Price: $125
   Tax: 6% _____

2. Price: $45
   Tax: 10% _____

3. Price: $240
   Tax: 15% _____

4. Price: $250
   Tax: 6% _____

5. Price: $24
   Tax: 10% _____

6. Price: $6
   Tax: 12.5% _____

7. Price: $24.90
   Tax: 20% _____

8. Price: $69
   Tax: 20% _____

9. Price: $648
   Tax: 6% _____

10. Price: $840
    Tax: 4.5% _____

11. Price: $19.95
    Tax: 20% _____

12. Price: $15
    Tax: 5% _____

Find each sale price.

13. Price: $52
    Discount: 25% _____

14. Price: $90
    Discount: 15% _____

15. Price: $80
    Discount: 20% _____

16. Price: $1,200
    Discount: 12% _____

17. Price: $36.50
    Discount: 40% _____

18. Price: $70
    Discount: 30% _____

19. Price: $17.80
    Discount: 10% _____

20. Price: $120
    Discount: 15% _____

21. Price: $250
    Discount: 18% _____

22. Price: $16.50
    Discount: 50% _____

Problem Solving
Solve.

23. A store discounts school supplies by 15%. The regular price of a pencil sharpener is $15.50. What is the sale price?

24. Another store discounts school supplies by 25%. The regular price of a notebook is $1.50. What is the sale price?

Use with Grade 6, Chapter 26, Lesson 1, pages 634–636.

Simple Interest

26-2 PRACTICE

Complete the chart. Use the formula $I = p \times r \times t$. Round interest to the nearest cent.

	Principal	Annual Rate of Interest	Time	Interest	Principal + Interest
1.	$2,000	10%	3 years		
2.	$15,000	14%	4 months		
3.	$3,000	16%	3 years		
4.	$720	12%	1.5 years		
5.	$980	8%	6 months		
6.	$1,250	18%	3 months		
7.	$7,000	15%	3.5 years		
8.	$25,000	15%	4 years		
9.	$3,400	9.5%	2 years		
10.	$500	5%	2 years		
11.	$1,000	8%	3 years		
12.	$2,500	6%	10 years		
13.	$1,500	9%	2 years		
14.	$1,080	12%	5 years		
15.	$2,000	8.5%	1 year		

Problem Solving
Solve.

16. Fred opened a savings account that earns 8% annual interest. How much will $2,000 earn in one year? _____

17. Harry borrows $2,500 to buy a car. The annual interest rate is 8%. Find the amount of interest he will pay for one year. _____

18. Joan deposits $600 in a savings account. The account pays 5% interest per year. How much money will be in the account at the end of 4 years? _____

19. Carol deposits $2,000 in a savings account that pays 6% interest annually. How much money is in the account at the end of 12 years? _____

Name _____

Circle Graphs

P 26-3 PRACTICE

Complete the table. Round to the nearest whole number.

Favorite Movies

Movie	Chosen by	Percent	Degrees
Drama	21 people	_____	_____
Comedy	87	_____	_____
Action	47	_____	_____
Science Fiction	19	_____	_____
Animated	38	_____	_____
Total	_____ people	_____	_____

Favorite Sports

Sport	Chosen by	Percent	Degrees
Baseball	41 people	_____	_____
Basketball	75	_____	_____
Soccer	31	_____	_____
Hockey	10	_____	_____
Gymnastics	14	_____	_____
Total	_____ people	_____	_____

Problem Solving
Solve.

1. Make a circle graph to show that 60% of the students in class have brown eyes, 25% have green eyes, and 15% have blue eyes.

2. What is the measure of the central angle for the part of the circle graph that represents 60%? _____

3. What is the measure of the central angle for the part that represents blue eyes? _____

Use with Grade 6, Chapter 26, Lesson 3, pages 640–642.

143

Name _____

Problem Solving: Strategy
Logical Reasoning

P 26-4 PRACTICE

Use a Venn diagram to solve.

1. There are 20 children in the Children's Choir that sing the soprano and alto parts. Thirteen children are sopranos and 4 children sing both soprano and alto. How many children sing alto but not soprano? What percent sing only alto?

   _____

2. Mr. Williams has 25 students working on the school paper and the school yearbook. Five of the students work on both the yearbook and the paper, while 12 students work on the paper only. What percent of the students work only on the yearbook?

   _____

3. Out of 165 dance students, 108 are ballet students. There are 95 jazz students. How many students take both ballet and jazz?

   _____

4. Out of 60 students, 12 have read only *James and the Giant Peach* by Roald Dahl and 28 have read both *James and the Giant Peach* and *Matilda* by Roald Dahl. Each student has read at least one of the books. What percent of the students have read only *Matilda*?

   _____

Mixed Strategy Review
Solve. Use any strategy.

5. For a soccer drill, the athletic director likes to have 3 soccer balls for each group of 7 players. If he is doing the drill with 84 players, how many soccer balls will he need?

   _____
   Strategy: _____

6. Ryan works in a shoe store. Today he sold 18 pairs of tennis shoes and 16 pairs of sandals. Seven people bought both tennis shoes and sandals. How many customers bought the shoes today?

   _____
   Strategy: _____

7. A train is traveling at a rate of 75 miles per hour. How many miles will the train travel in 3 hours 24 minutes?

   _____
   Strategy: _____

8. **Write a problem** which you could use logical reasoning to solve. Share it with others.

   _____
   _____

144

Use with Grade 6, Chapter 26, Lesson 4, pages 644–645.

Name _____

Probability

27-1 PRACTICE

Use the spinner shown to find each probability. Write it as a ratio, fraction, decimal, and percent.

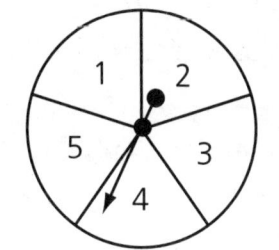

1. P(4)

_____

2. P(3)

_____

3. P(even number)

_____

4. P(number less than four)

_____

What if you select a marble from the bag without looking? Find each probability as a fraction in simplest form and as a percent. (You can see all the marbles in the bag.)

5. P(red) _____

6. P(blue) _____

7. P(green) _____

8. P(yellow) _____

Use the spinner shown to find each probability as a fraction in simplest form and as a decimal, rounded to the nearest thousandth.

9. P(6) _____

10. P(3) _____

11. P(number greater than 6) _____

12. P(odd number) _____

What if you mix the cards shown? Choose one without looking, record the outcome, and put it back. Find each probability as a fraction and as a percent, rounded to the nearest tenth of a percent.

13. P(T) _____

14. P(Q) _____

15. P(vowel) _____

16. P(consonant) _____

Use with Grade 6, Chapter 27, Lesson 1, pages 660–663.

145

Name _____

Explore Experimental Probability

27-2 PRACTICE

Solve.

1. Express as a percent the theoretical probability that you will get tails when you toss a coin. _____

   Then do 60 trials and find the experimental probability. Record the outcome of each toss in the table. Find the percent of times you got tails. _____

Heads	Tails

   How close were the theoretical probability and the experimental probability?

   _____

   _____

2. You are going to toss a number cube marked with the numbers 1 to 6. Express as a percent, rounded to the nearest whole number, the theoretical probability that you will get a number greater than 2. _____

   Then do 50 trials and find the experimental probability. Record the outcome of each toss in the table. Find the percent of times you got 3, 4, 5, or 6. _____

1 or 2	3, 4, 5, or 6

   How close were the theoretical probability and the experimental probability?

   _____

   _____

Use with Grade 6, Chapter 27, Lesson 2, pages 664–665.

Make Predictions

27-3 Practice

Predict the number of times you would get each color if you do this experiment 600 times: Take a marble from the bag without looking, write down the color, and put it back.

1. red _____   2. yellow _____   3. blue _____

Mary drew blocks from a bag without looking, recorded each color, and put each back before selecting another. Use data from the table to predict how many blocks of each color are in the bag, if the bag contains 200 blocks.

Color	Number Selected
Red	5
Blue	15
Green	10
Yellow	5
Black	2
Purple	8
Orange	4
White	1

4. purple _____   5. blue _____

6. green _____   7. yellow _____

8. black _____   9. orange _____

10. red _____   11. white _____

Use data from the random survey responses to predict the total number of people who would give each answer.

Which do you prefer as an after-school sport?	
Sport	Number of Students
Soccer	12
Softball	32
Lacrosse	4
Volleyball	2
Total surveyed: 50 students out of 850 students	

12. soccer _____

13. softball _____

14. lacrosse _____

15. volleyball _____

Use with Grade 6, Chapter 27, Lesson 3, pages 666–668.

Name _____

Mutually Exclusive Events

P 27-4 PRACTICE

What if you select a card at random? Find each probability as a fraction in simplest form.

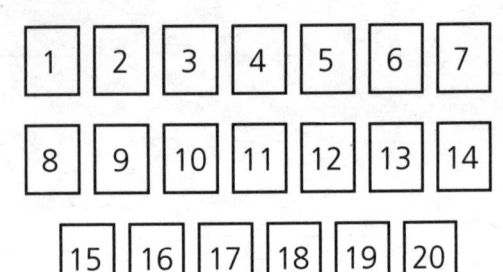

[1] [2] [3] [4] [5] [6] [7]
[8] [9] [10] [11] [12] [13] [14]
[15] [16] [17] [18] [19] [20]

1. P(number less than 2 or greater than 11) _____

2. P(number less than 8 or greater than 9) _____

3. P(number less than 3 or greater than 10) _____

4. P(8 or 9) _____

5. P(number less than 12 or greater than 13) _____

6. P(not 11) _____

7. P(number less than 8 or greater than 14) _____

Find each probability. Show that $P(A \text{ or } B) = P(A) + P(B)$.
Experiment: Select a card from 25 cards at random. The cards are numbered 1–25.

8. P(number less than 4 or greater than 20)

_____

_____

9. P(number less than 11 or greater than 14)

_____

_____

10. P(number less than 10 or greater than 18)

_____

_____

Name _____

Problem Solving: Skill
Check for Reasonableness

27-5 PRACTICE

Estimate to check whether each claim is reasonable, then solve. Use data from the table for problems 1–2.

Activity	Number of Students
Basketball	41
Volleyball	10
Racquetball	7
Swimming	22

A random sample of 80 students from the 200 students that use a recreation center answered the question "Which activity do you enjoy the most: basketball, volleyball, racquetball, or swimming?" Cameron made a table of the responses.

1. Cameron says that 27% of the sample chose swimming. Is his calculation reasonable? Explain.

_____

_____

2. Cameron predicts that 10 students at the center enjoy volleyball the most. Is his prediction reasonable? Explain.

_____

_____

Choose the correct answer.

In a random sample of 40 students asked to name their favorite subject, 16 said history. Therefore, it is likely that 120 students in a group of 300 would have the same response.

3. Which of the following is true?

A $\frac{16}{40} = \frac{120}{300}$

B $\frac{16}{120} = 40\%$

C $0.04 = \frac{16}{40}$

D $\frac{15}{120} = \frac{40}{300}$

4. If an answer is reasonable, then it

F must be an estimate.

G should be a percent.

H should not be a sum.

J makes sense.

Mixed Strategy Review

5. Lauren makes picture frames using small shells in a 1 to 6 ratio of white shells to black. If she uses 84 shells per frame, how many shells of each color will she need?

_____

6. Erik wants to wallpaper a wall. The wall is 11.5 feet wide and 8 feet high. How much paper does he need?

_____

Use with Grade 6, Chapter 27, Lesson 5, pages 674–675.

149

Compound Events

P 28-1 PRACTICE

Make a tree diagram to find the number of possible outcomes.

1. spinning each spinner once _____

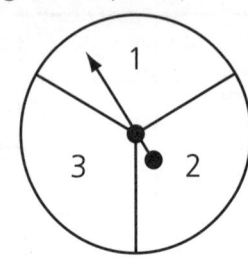

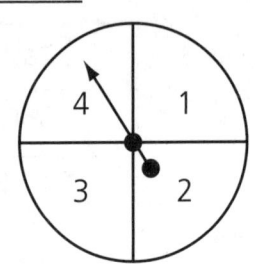

2. tossing a coin and tossing a 7–12 number cube _____

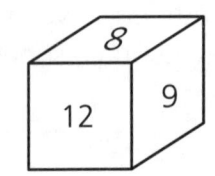

3. tossing a 1–6 number cube and spinning the spinner _____

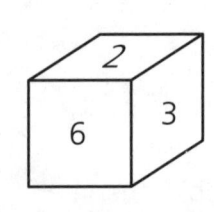

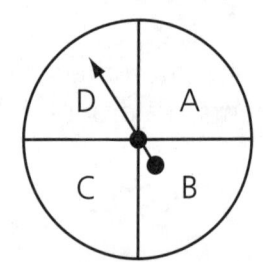

4. selecting a skirt from a choice of blue, black, grey, or tan, and a blouse from a choice of blue, green, red, orange, or yellow _____

Find the total number of possible outcomes.

5. selecting a pitcher and catcher from 5 pitchers and 3 catchers _____

6. selecting a car from 8 models and 7 colors _____

7. ordering a sundae by choosing 1 out of 3 toppings and 1 out of 12 flavors of ice cream _____

8. selecting a pair of glasses from 7 colors and 14 styles _____

9. selecting a pair of shoes from a choice of brown, black, or white, and a pair of socks from a choice of green, black, brown, navy, or white _____

10. ordering a pizza by choosing 1 out of 4 kinds of crust and 1 out of 16 toppings _____

150 Use with Grade 6, Chapter 28, Lesson 1, pages 680–683.

Name _____

Problem Solving: Strategy
Make an Organized List

P 28-2 PRACTICE

Make an organized list to solve.

1. How many different ways can two CDs be chosen from a collection of five CDs?

   _____

2. Four friends are running on a track. How many different ways can the four friends run together in pairs?

   _____

3. Jules, Miranda, and Enrique are waiting to have their names called. What is the probability that the names will be called in the order Jules, Miranda, and Enrique?

   _____

4. Dena has 5 ribbons. They are blue, red, yellow, purple, and green. What is the probability that she will randomly choose a blue ribbon and a green ribbon at the same time?

   _____

Mixed Strategy Review
Solve. Use any strategy.

5. The cost of repairing a CD player is $50 plus $45 per hour for labor. If the bill is $230, how many hours of labor were needed?

   _____

   Strategy: _____

6. Corey had 6 rolls of film developed. Some rolls had 24 pictures and some had 36 pictures. He took 156 pictures in all. How many rolls of each type did Corey use?

   _____

   Strategy: _____

7. Eric is buying a pair of jeans on sale for 15% off the regular price. The regular price of the jeans is $35. If the tax is 6.5%, what is the total cost of the jeans?

   _____

   Strategy: _____

8. Write a problem for which you could make a list to solve. Share it with others.

   _____
   _____
   _____

Use with Grade 6, Chapter 28, Lesson 2, pages 684–685.

Name _____

Explore Independent and Dependent Events   **P** 28-3 PRACTICE

Tell whether the two events are independent or dependent.

1. You toss a nickel, and then you toss a dime. _____

2. You draw a card at random, then do not replace it, and then you draw another card at random. _____

3. You grab a sock from the dryer, and then you grab another sock from the dryer. _____

4. You draw a card at random, then replace the card, and then you draw another card. _____

5. You open your math book to a page, then close it, and then you open it again. _____

6. A drawer contains 4 red marbles and 6 blue marbles. You draw a marble at random, then replace it, and then you draw another marble. _____

7. Choose a bracelet and put it on. Then choose another bracelet. _____

8. Pick one flower from a garden, and then pick another flower. _____

9. Choose a pair of pants from a closet of pants. Then choose a shirt from a drawer of shirts. _____

10. In a hardware store, choose a hammer to buy and then choose a color of paint. _____

Name_____

Independent and Dependent Events

P 28-4 PRACTICE

Toss each 1–6 number cube once. Find the probability as a ratio and as a fraction in simplest form.

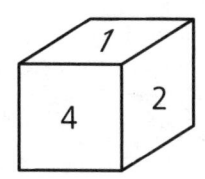

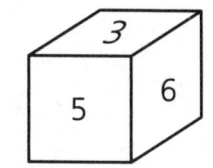

1. P(1 and 6) _____
2. P(3 and even) _____
3. P(odd and even) _____
4. P(6 and 4) _____
5. P(5 and 5) _____
6. P(6 and odd) _____

A card is selected from the bag without looking, the letter recorded, and the card is not replaced. A second card is selected without looking and the letter recorded. Find each probability as a fraction in simplest form and as a decimal rounded to the nearest thousandth.

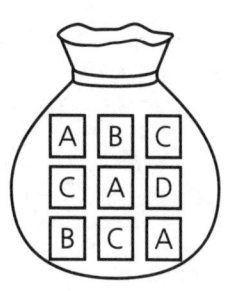

7. P(A and B) _____
8. P(A and A) _____
9. P(consonant and A) _____
10. P(vowel and B) _____
11. P(B and C) _____
12. P(C and D) _____

Make a tree diagram for the experiment. List the probability of each outcome.

13. **Experiment** Select two marbles at random from a bag of 3 blue and 2 yellow marbles. Do not replace the first marble before selecting the second.

_____

Problem Solving
Solve.

14. A wallet contains three $1 bills, four $5 bills, and five $20 bills. Two bills are selected without the first selected bill being replaced. Find the probability of selecting a $5 bill and then a $1 bill.

15. A purse contains five dimes, two nickels, and three pennies. Two coins are selected without the first selected coin being replaced. Find the probability of selecting a dime and then a penny.

Use with Grade 6, Chapter 28, Lesson 4, pages 688–690.

153

Summer Skills Refresher

Summer Skills

Planning a Picnic

Many beaches on the East Coast have boardwalks along their edges. Boardwalks are like outdoor shopping malls where people can stroll, buy food, clothes, and toys, and play games. Families can plan an entire day around the beach and the boardwalk.

Family Fun

Some beaches charge money to allow you on to the beach. At one beach, tags for adults are $7.50 and tags for children ages 3–12 are $3.50. Children under 3 are free. You can also rent beach umbrellas for $4.00 an hour.

1. Maya's family went to the beach for the day. There are 5 people in Maya's family: her mom and dad, one brother who is 3 months old, another brother who is 15, and Maya, who is 12. How much will beach tags cost Maya's family?

   _____

2. Maya's family decided to rent an umbrella. They rented the umbrella for 4 hours. How much did they pay for the umbrella?

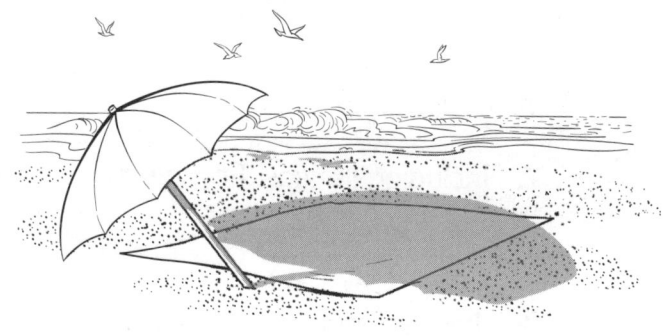

   _____

3. Maya's father bought lunch on the boardwalk for the family. He bought four hot dogs for $2.75 each and two large orders of fries for $2.50 each. How much did Maya's father spend?

   _____

4. Maya's father also purchased four sodas for $1.25 each. How much did he spend for all of the food and the drinks?

   _____

Answers: 1. $26.00; 2. $16.00; 3. $16.00; 4. $21.00

Summer Skills 157

Julian's New Job

When Maya's father bought lunch, he noticed there was a Help Wanted sign in the window. He encouraged Maya's older brother, Julian, to apply for a job at the restaurant. Julian filled out an application and got a job working 3 days a week at the restaurant.

5. Julian's salary is $5.85 an hour, plus any tips that customers give him. The first week he was scheduled to work three 6-hour shifts. How much did Julian earn (without tips) for the hours he worked?

   _____

6. Julian earned a total of $178.63 his first week. How much did he earn in tips?

   _____

7. Including the tips, how much money did Julian make per hour?

   _____

8. Julian went to the movies. He had only his tip money with him. Julian spent $7.50 for his movie ticket, $2.75 for popcorn, and $2.25 for a soda. How much money did Julian have left?

   _____

9. Julian worked overtime on Saturday night. Overtime pay is $6.25 an hour. He earned $31.25 without tips. How many hours did he work?

   _____

Answers: 5. $105.30; 6. $73.33; 7. $9.92; 8. $60.83; 9. 5 hours

Summer Skills

What a Grand Trip!

The Grand Canyon in Arizona is a popular vacation destination. Visitors enjoy train and mule rides, hiking, and even river rafting. Tourists need to book reservations well in advance.

Flying High

A helicopter company flies tourists around the Grand Canyon so they may get a bird's eye view. Each tourist must provide his or her weight to the company when making reservations. The company needs the weights to plan a balanced seating arrangement for the helicopter.

1. One family, from Europe, submitted their weights in kilograms. The 5 passengers weighed: 50 kg, 60 kg, 70 kg, 80 kg, and 90 kg. If 1 kilogram is equal to 2.2 pounds, how many pounds did each passenger weigh?

2. Design a seating plan. Place 2 people on one side and 3 people on the opposite side with less than a 25-pound difference between the two sides.

3. How many pounds did the tourists weigh in all?

4. The maximum weight allowable for passengers in the helicopter is half a ton. The helicopter company wanted to fill the sixth passenger seat with a photographer. What was the greatest number of pounds the photographer could weigh?

Answers: **1.** 110 lbs, 132 lbs, 154 lbs, 176 lbs, 198 lbs; **2.** First side: 110, 132, 154, 396 in total; Second Side: 198, 176, 374 in total; **3.** 770 pounds; **4.** 230 pounds

Summer Skills 159

Riding the River

Another company offers rafting trips through the canyon. The trips are advertised as half-day tours that use motorized rafts.

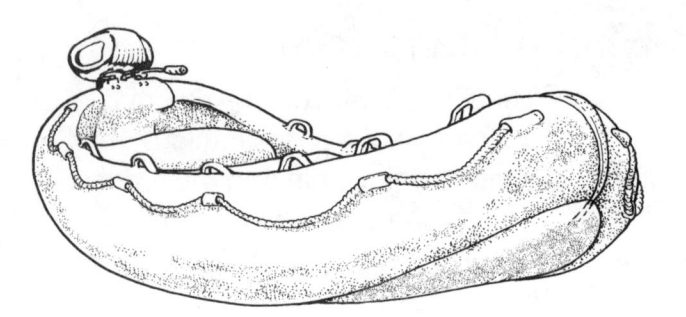

5. The tours start at 8:30 A.M. and 12:30 P.M. The Weiss family arrived Friday morning at the time shown on the clock below. How long will they have to wait until the next tour departs?

6. The tour that leaves at 12:30 travels for 2 hours and 15 minutes and then stops for 15 minutes at Dangling Rope Marina. What time will the raft leave the marina?

7. The raft stops a second time at Rainbow Bridge. They spend 45 minutes eating a snack. They leave the bridge at the time shown on the clock below. How long did it take the raft to travel from Dangling Rope Marina to Rainbow Bridge?

Answers: 5. 2 hours and 8 minutes; 6. 3:00; 7. 1 hour and 10 minutes

Summer Skills

Growing a Garden

Marisa and her mother are planning to plant a garden in their backyard. Before they can begin planting, they have to tear up the grass and get the soil ready.

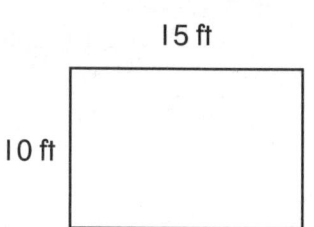

1. Marisa's mother marked the grass in the shape of a rectangle to show the outline of the garden. The length of the garden measures 15 feet and the width measures 10 feet. What will the area of the garden be?

   _____

2. Marisa's mother wants to place a fence around the perimeter of the garden. The fence will be 6 feet tall. What is the total square footage of the fence they will need?

   _____

3. Marisa is going to fertilize the soil in the garden. Each bag of fertilizer will cover 5 square feet. How many bags does Marisa need to cover the entire garden?

   _____

4. They decide to put mulch around the outside of the garden fence to create a border around the entire garden. If they make the border 1 foot wide on all sides, what is the total perimeter of the garden, including the mulch?

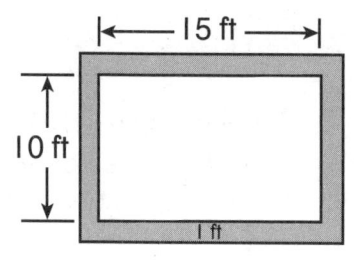

   _____

Answers: 1. 150 sq. ft; 2. 300 sq. ft; 3. 30 bags; 4. 58 ft

More Flowers for Marisa

While Marisa and her mother are preparing the garden, Marisa's brother, Manny, is making window boxes to hold flowers in front of the house.

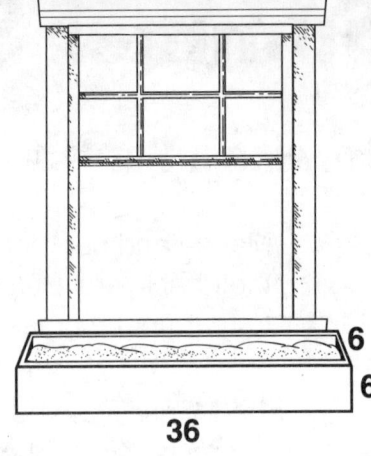

5. Each window box will be 6 inches wide, 6 inches high, and 36 inches long. The boxes will all be filled with soil. What is the volume of each box to be filled?

6. The outside surface area of the window boxes will need to be painted. What is the surface area that will need to be painted on each box? Remember: There is no top on the box.

7. If there are 4 boxes in all, what is the total surface area that needs to be painted?

8. Marisa's mom thinks the window boxes are too large. She asks Marisa's dad to cut the tops down so the boxes measure 36 inches × 4 inches × 6 inches. What is the difference in volume between the old boxes and the new boxes?

Answers: 5. 1,296 cu. in.; 6. 720 sq. in.; 7. 2,880 sq. in.; 8. 432 cu. in.

Summer Skills

Buying a Bike

Janeen was working two summer jobs in order to earn enough money to buy herself a new bike. She delivered papers in the morning and worked at the market in the afternoon.

1. Janeen earned $10.00 more each day at the market than she did delivering papers. She earned $30.00 a day at the market. Let d equal the amount of money she earned delivering papers each day. Write an equation that represents how much Janeen earned delivering papers each day.

   _____

2. Solve the equation from problem 1 to find the amount of money Janeen earned delivering papers each day.

   _____

3. Let w equal the amount of money Janeen earns delivering papers for 7 days. Write and solve an equation to find how much money Janeen earns delivering papers for 7 days.

   _____

4. Janeen works at the market only 4 days a week. Let m equal the amount of money Janeen earns working at the market each week. Write and solve an equation to find how much money Janeen earns working at the market each week.

   _____

Answers: 1. $d + $10.00 = $30.00; 2. $20.00 each day; 3. 7($20.00) = w, w = $140.00; 4. 4($30.00) = m, m = $120.00

Summer Skills 163

Up, Up and Away

Every year, towns all over the country host balloon festivals. Most offer massive lift-offs at dawn and dusk, when hundreds of hot air balloons take off within several minutes of each other.

5. Sarah and her 4 immediate family members met up with her extended family at a balloon festival near her house. There was a total of *x* family members in attendance, including Sarah and her immediate family. Write an expression to represent the number of family members, not including Sarah or her immediate family.

6. On the second night of the festival, 148 balloons floated into the air. That was 33 more than on the first night. How many balloons lifted off the first night?

7. Balloon rides cost $75.00 plus $10.00 per person. Sarah and her immediate family decided to go on a ride together. Write an equation to show the cost of the balloon ride. Solve the equation.

8. Sarah's family was able to purchase pictures of their balloon flight. They bought 3 large pictures for $7.00 each and 2 small pictures for $4.00 each. Write and solve an equation that shows how much they spent on pictures.

Answers: **5.** $x - 5$; **6.** 115; **7.** $C = \$75.00 + 5(\$10.00)$, $C = \$125.00$; **8.** $C = 3(\$7.00) + 2(\$4.00)$, $C = \$29.00$

Summer Skills

Summer Fun in Town

The Bridgewater Recreation Department runs a summer program for children ages 3–12. The Town Council wanted to make certain that it provided programs for students of different ages. In the spring, the council surveyed students of various ages to plan summer programs.

1. The council wanted to get a representative sample of the students who would be participating in the activities. How could members of the council create a survey that would produce a representative sample?

   _____

   _____

2. Results of the survey are shown in the table below. Use the results for Girls Ages 3–6 to complete the bar graph below.

Activity	Children Ages 3–6		Children Ages 7–12	
	Boys	Girls	Boys	Girls
Sports	10	17	15	8
Computers	8	8	14	10
Arts and Crafts	10	15	5	18
Swimming and Boating	15	2	10	10
Science Enrichment	7	8	6	4
Total Surveyed	50	50	50	50

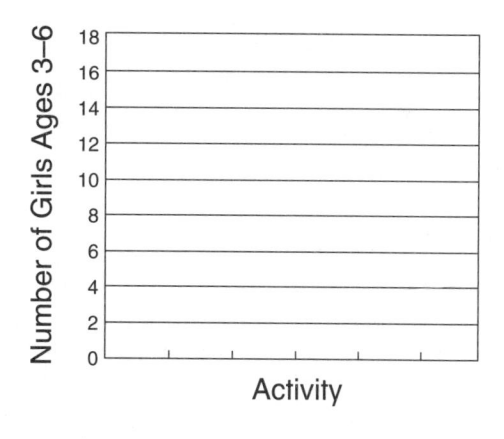

Answers: 1. Possible answer: They could ask 5 boys and 5 girls in each age group. 2. Check graph

Summer Skills 165

The Science Club

After seeing results of the survey, the Bridgewater Recreation Department decided to have one session of science activities during its summer program. Students made rain collectors to collect and measure the rainfall.

3. The students left their rain collectors outside and at the end of each week, they measured the rain that had fallen into them. The chart below shows the rainfall amounts.

   Make a line graph of the rainfall amounts. Be sure to label your graph.

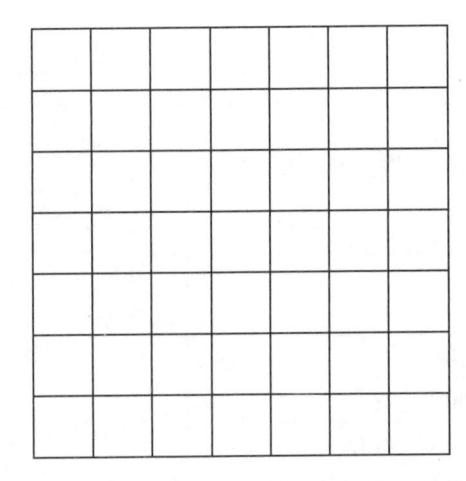

Week 1	3 inches
Week 2	2 inches
Week 3	3 inches
Week 4	5 inches
Week 5	1 inch

4. What was the mean amount of rain that fell during the weeks the students collected rain?

   _____

5. What is the mode of the data? _____

6. The students also recorded the high temperature every day for 2 weeks. Find the mean temperature during the two-week time period (rounded to the nearest tenth).
   78°, 83°, 79°, 90°, 86°, 88°, 83°, 78°, 78°, 80°, 83°, 85°, 85°, 87°

   _____

Answers: 3. Check graph; 4. 2.8 in.; 5. 3 in.; 6. 83.1°